Engineering Materials

This series provides topical information on innovative, structural and functional materials and composites with applications in optical, electrical, mechanical, civil, aeronautical, medical, bio- and nano-engineering. The individual volumes are complete, comprehensive monographs covering the structure, properties, manufacturing process and applications of these materials. This multidisciplinary series is devoted to professionals, students and all those interested in the latest developments in the Materials Science field, that look for a carefully selected collection of high quality review articles on their respective field of expertise.

Indexed at Compendex (2021) and Scopus (2022)

Bhasha Sharma · Vijay Chaudhary ·
Chanchal Ahlawat · James Njuguna
Editors

AI-Driven Polymer Design

Structure, Properties, and Applications

Springer

Editors
Bhasha Sharma
Department of Chemistry
Deen Dayal Upadhyaya College
University of Delhi
New Delhi, Delhi, India

Chanchal Ahlawat
School of Computer Science Engineering and Technology
Bennett University
Greater Noida, Uttar Pradesh, India

Vijay Chaudhary
Department of Mechanical Engineering
Amity University Uttar Pradesh
Noida, Uttar Pradesh, India

James Njuguna
School of Engineering
Robert Gordon University
Aberdeen, UK

ISSN 1612-1317 ISSN 1868-1212 (electronic)
Engineering Materials
ISBN 978-981-95-8352-2 ISBN 978-981-95-8353-9 (eBook)
https://doi.org/10.1007/978-981-95-8353-9

This Springer imprint is published by the registered company Springer Nature Singapore Pte Ltd.
The registered company address is: 152 Beach Road, #21-01/04 Gateway East, Singapore 189721, Singapore

Preface

This book, *AI Driven Polymer Design: Structure, Properties, and Applications*, will provide a comprehensive exploration of how artificial intelligence (AI) is transforming the design, analysis, and application of polymers in modern science and industry. AI technologies, including machine learning (ML) and deep learning, have revolutionized the discovery, prediction, and optimization of polymers, enabling the rapid development of materials with specific properties suited for various applications such as healthcare, energy, and electronics.

The book will cover the role of AI in understanding the relationship between the molecular structure of polymers and their physical, chemical, and mechanical properties. By integrating advanced AI models, the book will demonstrate how new polymeric materials can be designed more efficiently, with tailored properties for targeted applications. This work will be both theoretical and practical, offering insights into AI algorithms, predictive modeling, polymer property analysis, and case studies.

The book will be valuable for materials scientists, chemists, AI researchers, and industry professionals seeking to leverage AI in polymer science and material design.

The book aims to:

- Provide a detailed introduction to AI-driven polymer design, from molecular structure prediction to property optimization.
- Explain the relationship between polymer structure, properties, and their applications.
- Showcase the use of AI for predictive modeling in polymer science, enhancing the understanding of mechanical, thermal, and chemical behaviors.
- Highlight the real-world applications of AI-optimized polymers across multiple industries such as biomedical engineering, aerospace, and electronics.
- Discuss the challenges, opportunities, and future prospects for AI in polymer research and development.

AI-Driven Polymer Design: Structure, Properties, and Applications, will provide an essential, up-to-date resource for researchers and industry professionals in materials science and polymer chemistry. By integrating artificial intelligence with polymer design, this book will offer both theoretical insights and practical tools for advancing the field of polymer science, demonstrating how AI can shape the future of materials discovery and innovation.

New Delhi, India — Bhasha Sharma
Noida, India — Vijay Chaudhary
Greater Noida, India — Chanchal Ahlawat
Aberdeen, UK — James Njuguna

Contents

Introduction to Artificial Intelligence in Plastics and Polymeric Products Design

Bhuvi Misra, Pragya Sharma, V. P. Sharma, Bhasha Sharma, Vijay Chaudhary, and James Njuguna

Abstract Polymers are versatile and significant components of the modern material industry. They offer unique properties and are indispensable in varied sectors, including manufacturing, healthcare and infrastructure. Plastics, derivative of these polymers, are omnipresent and contribute expansively to everyday life and industrial sectors. They are highly ubiquitous due to their cost-effective, malleable, and durable design. The polymer industry has heavily relied on Edisonian methods to innovate products like plastic. These approaches were labour-intensive, expensive and required prolonged research. While these methods have led to revolutionary innovations, they are resource-intensive and limited in their ability to generate complex polymeric designs. The integration of Artificial Intelligence into polymeric product designing has marked a new era in the polymer industry. The AI-generated data-driven methods help to bridge the gap between theoretical models of polymers and their empirical observations. They assist in material design and discovery, property prediction and optimisation, manufacturing processes, and categorisation, while also uncovering insights for sustainable polymer generation. These methodologies encompass various Machine Learning and Deep Learning. A brief insight into the basics of artificial intelligence, data-driven models and polymeric designing has

B. Misra · V. P. Sharma (✉)
CSIR-Indian Institute of Toxicology Research (CSIR-IITR), Lucknow, UP, India
e-mail: vpsitrc@rediffmail.com

P. Sharma
National Artist & Dastan Go On Environmental Issues, Lucknow, UP, India

B. Sharma
Deen Dayal Upadhyaya College, University of Delhi, Delhi, India
e-mail: sharmabhasha@gmail.com

V. Chaudhary
Department of Mechanical Engineering, Amity School of Engineering & Technology, Amity University Uttar Pradesh, Noida, India
e-mail: vijaychaudhary111@gmail.com

J. Njuguna
Research and Innovation, The National Subsea Centre (NSC) and Faculty, Robert Gordon University (RGU), Aberdeen, UK
e-mail: j.njuguna@rgu.ac.uk

B. Sharma et al. (eds.), *AI-Driven Polymer Design*, Engineering Materials,
https://doi.org/10.1007/978-981-95-8353-9_1

been introduced and discussed in this chapter. These methods not only modernize design processing but also enable simulation and real-time monitoring of the polymer products. By overcoming the shortcomings of empirical methods through advanced computational techniques, artificial intelligence is aiming to overcome longstanding challenges associated with polymer formulation for creating sustainable polymer products.

Keywords Artificial intelligence · Deep learning · Environment · Evolutionary algorithms · Machine learning · Polymer design · Sustainability

1 Introduction

Polymers and plastics are essential elements of contemporary industries and have proliferated during the twentieth century. They deliver flexible material solutions at reasonable rates for an array of businesses, including consumer goods, packaging, healthcare, automotive, and aerospace. This interaction between the manufacturing sector and polymer science is essential to developing innovative substances that meet the ever-changing requirements of the contemporary world [1–3].

Because polymers allow products to have unique qualities for certain applications they have completely changed the fast-moving consumer goods (FMCG) market. Because polymer chemists have perfect control over the molecular architecture of these compounds, they may create materials with a wide range of qualities, such as flexibility, durability, resistance to temperature, etc. For example, thermoplastics are acclaimed for being recyclable since they can be remelted and transformed. conditions. In recent years, researchers have integrated advanced simulation tools and modelling techniques to predict a product's behaviour under various conditions. While these methods have delivered numerous innovations, they are often limited by sheer amount of time and data required, and, nature and complexity of polymer behaviour and development [4–6] .

Artificial Intelligence (AI) has dramatically expanded its capabilities beyond the realms of its promised prototypes. Its incorporation in the design and manufacturing methods has effectuated substantial transformation and given rise to a new interdisciplinary field known as **materials informatics (MI)**. MI utilises AI-driven methods to run simulations and predict outcomes in manufacturing industries. The plastic industry has also begun to harness AI's potential despite its traditionally slow adaptation towards technological shifts [7].

Historical traditions of designing relied heavily on empirical and physics-based approaches. While these were grounded in principles, they were constrained by high experimental cost, lengthy development cycle and difficulty in management of complex and multivariable designs. Many of such challenges in material design are being overcome by AI based programs. It offers an appealing solution with its ability to model complex system. In scenarios where real-world experimentations are time-consuming or expensive, AI-based simulations offer a favourable alternative.

Additionally, these models provide real-time monitoring over production to ensure quality of product under variable conditions [8].

Artificial intelligence (AI), through improving material development, automating production, and promoting sustainability measures, is revolutionizing the plastic sector. AI-driven algorithms predict material behavior by analyzing massive datasets, leading to the development of more durable, long-lasting, and ecologically friendly polymers. Manufacturing optimisation increases production efficiency by predicting and managing several aspects, reducing waste, and guaranteeing high-quality output. By foreseeing possible flaws in machinery, predictive maintenance with IoT sensors driven by artificial intelligence may decrease downtime and boost operational efficiency. Quality and assurance activities aid in the detection of systems that use high-resolution cameras in conjunction with machine learning algorithms to identify inconsistencies in products, ensuring that strict quality requirements are revolutionising the recycling process by accurately sorting and processing different plastics. As the recycling industry transitions to a circular economy, this helps to detect systems that use cameras with excellent resolution along with machine learning algorithms to identify inconsistencies in products. This supports the recycling sector as it moves to a circular economy [9–11].

2 Fundamentals of Artificial Intelligence

The computational technology known as artificial intelligence (AI) permits computers and other technologies to promote learning, comprehension, creativity, problem-solving, and decision-making with a minimal amount of human involvement. It learns from data, identifies patterns, and formulates judgments by simulating human mental processes. AI-supported applications see and learn from human language, offer recommendations, and take independent action based on evaluations of past experiences. To put it simply, intelligent technology is a collection of ideas and technologies. It has developed over seven decades and continues to become bigger and bigger. Machine learning, deep learning, and generative AI are more sophisticated kinds of artificial intelligence, whereas artificial intelligence (AI) is the general term for intelligence similar to human beings expressed by computational machines. AI is a more sophisticated version of the same. In recent and upcoming years, generative AI (gen AI) is taking precedence over other forms of AI on which it is built, i.e., machine learning and deep learning [12].

Machine learning (ML) is the process of using previous data to anticipate outcomes and create models. Through the use of a wide variety of algorithms, it can learn from and infer prior data without requiring special training, and as it is provided with additional information, its performance improves. Machine learning methods such as logistic regression, the use of support vector machines, random forest modeling, k-nearest neighbors, logistic regression, clustering, and decision trees are all made to solve different kinds of problems with the environment [13].

To predict accurate results, supervised machine learning uses data that has been labeled with output and backed by training examples. This approach aims to teach the model to map input and output with the objective to forecast new and previously unidentified information. Image identification, fraud detection, and risk assessment are common applications of training under supervision. The artificial neural network (ANN) is the most commonly employed supervised artificial intelligence system. The human nervous system, where networks are linked by layers of nodes that resemble neurons, is the model for this method. An ANN may include one or more intermediary layers, referred to simply as hidden layers, in addition to its single input and output layers. artificial neural networks (ANN are capable of finding patterns in vast, sophisticated data sets. A learning algorithm having a hierarchy of trees as its framework is called a decision tree (DT).It is a non-parametric method used for predicting potential outcomes. Root nodes, internal nodes, and leaf nodes make up this top-to-bottom construction process. The complete dataset needed to train the model is contained in the root node. The feature that best characterizes the dataset is used to divide it. This minimizes the possibility of split sets overlapping. This decision node is further branched into the attribute of that set in order to further classify. Until a leaf appears, signifying the last prediction of that data, the whole procedure is repeated [14–16]. This machine learning method is favored because it is very interpretable when dealing with continuous and category-based information for classification and regression analyses.

Regression techniques that use associations between actual or continuous data to predict variables include the use of logistic regression, random forest, and linear regression. A measurement of one variable is used in linear regression to forecast the value of another. It is a relatively easy model which offers a mathematical formula for making predictions that is simple to understand. Organisations employ linear regression, which makes use of vast amounts of data and aids in better decision-making as a replacement to depending on experience and conjecture. On the other hand, the estimated likelihood of the assessed category set of independent variables being present is provided by logistic regression. Since this model provides a likely answer, the end result falls in the range of 0 to 1.A learning methodology called random forest uses selected samples from the datasets to create an ensemble of various decision trees. For each tree, the algorithm selects data at random from the set. Since each tree is trained using a distinct subset of the data, they are all distinct from the others. As a result, each tree develops unique predictions on the basis of the data it has been educated on, which add contribute to the final prediction [5–8].

As a classification algorithm, k-nearest neighbors (kNNs) allocation untrained evidence to the class that has the most comparable labeled samples. It is predicated on the idea that nearby data sets with comparable variables are accessible. The number of adjacent measurements selected for the kNN method is determined by the parameter "k." The results of kNN categorization will be influenced by the selection of "k." A high "k" value increases the number of small but significant patterns that are missed, even though it lessens the effect of error-induced variance. Similar to this, a tiny "k" the number makes the model extremely responsive to each item of information entered and could result in unpredictable predictions due to mismatched data and high variability [17, 18].

The square root of the training dataset's observed count is what should be used to get the "k" value. Finding a hyperplane that clearly divides data of various kinds is the main idea. This algorithm's straightforward methodology and operation made it suitable for disease prediction. Support vector regression, often known as SVR, is a type of regression job that may be performed using Support Vector Machines (SVMs), which are mainly used for classifications. By building a non-linear model with "kernel" mapping—which is known to rely on intricate database mapping into higher dimensions in an arrangement that facilitates data separation—it produces an improvement in the outcome. Unlike supervised learning, unsupervised machine learning infers information from data sets with labels and enables exploratory predictions. The method most frequently employed for unsupervised learning is cluster analysis, also known as clustering. It works by classifying data based upon comparable values and integrating several algorithms into one group. Data scientists may find item correlations in massive datasets with the assistance of these algorithms, delivering output through comparison. To improve learning, DNNs are specialized ANNs that have three to hundreds of hidden layers [19, 20].

DNNs are capable of extracting data from immense unlabeled and unstructured datasets thanks to multilayering, and then use that data to learn and make predictions. For instance, it may forecast stock market shifts through examining past stock data using several hidden layers. Because DNNs don't need human intervention, they facilitate machine learning on a greater scale. The processing of natural languages (NLP), image recognition, and other jobs requiring sophisticated algorithms for rapid and accurate results represent just a few of the sophisticated applications of AI which have been improving every aspect of our lives [7, 8].

Convolutional neural networks (CNNs) are a type of DNN primarily used to analyse and learn from visual datasets. For instance, identifying a human in an image or video by detecting shapes, edges, and textures. One kind of DNN that is mostly utilized for learning and analyzing information from visual datasets is the convolutional neural network (CNN). For example, using textures, edges, and shapes to identify an individual in a video or image. Self-supervised learning, mixing both supervised and unsupervised learning, and applying knowledge from one dataset to comprehend another related dataset are all made possible by the intricate nature of deep learning. A related type of deep learning is becoming more popular. In addition to reward functions, reinforcement learning, additionally referred to as reinforcement learning with human feedback (RLHF), learns by making missteps. Instead than gathering data from layers that remain concealed, it relies on an independent agent to make recommendations [3–5].

One kind of DNN that is mostly utilized for learning and analyzing information from visual datasets is the convolutional neural network (CNN). For example, using textures, edges, and shapes to identify an individual in a video or image. The complexity of deep learning enables self-supervised learning, combining supervised learning with unsupervised learning, and transferring knowledge gained from one dataset to understand another related dataset. A similar form of deep learning is emerging as new. **Reinforcement learning**, or **reinforcement learning with human feedback (RLHF)**, learns by trial and error, alongside reward functions. It depends on an autonomous agent for decision-making rather than extracting information through hidden layers [7–9].

Robots and self-driving cars are examples of autonomous machines that make decisions on their own without direct human assistance by interacting with their surroundings. In situations that are unpredictability and variable, it employs a sequential decision-making method. Unlike supervised learning, RLHF mimics real-world learning techniques to reward knowledge instead of using clearly labeled both right and wrong responses or outputs [6–8].

Generative AI, also called "gen AI," are deep learning models that generate original context through complex learning algorithms, such as images, videos, audio, text, and more, in response to the user's input/prompt. A foundation model is created through time-consuming, expensive, compute-intensive training to create a successful and functional gen AI. This model is subjected to a large amount of relevant data, which are raw, unlabelled and unstructured. The training results in a heavily complex neural network of the encoded data in response to datasets. This results in a complete foundation model that is further tuned for the generation of specific content. The model can be tuned by exposing the model with specific labelled data questions that the application will likely receive and correct answers in the desired format. Alternatively, the tuning can be done through RLHF, where users' simple feedback/input prompts to the chatbot can generate the desired response. Finally, the tuned model is accessed over its outputs and tuned regularly (as often as a week) for relevance and accuracy. In contrast, the foundation model can remain untouched for a year or more [4–6].

Recent research and development of computational methods and algorithms have led to a heuristic search approach called **Evolutionary algorithms (EA)**. This term is used to describe systems which are based on Darwinian evolution and solve iterative problems based on population. Some of the main algorithms of EA are:

- **Genetic algorithm (GA)**: This is one of the oldest and most well-known techniques based on Darwinian evolution. It is a universal tool for optimisation and mimics natural processes to generate potential results. GA operates on a population where each individual (referred as a chromosome) holds a potential solution to the problem. These individuals are evaluated based on how suitable they are for the objective function, which is known as the fitness of that individual. The algorithm optimises the population by using operators like selection (new population based on fitness value), crossover (typical similarities between two individuals selected to crossover) and mutation (random change in values of genes) to filter solutions.
- **Genetic programming (GP)**: This is a relatively new and specialised form of GA. It uses modified operators to generate specific solutions. In GP, the solutions are coded as trees, compared to linear chromosomes in GA, and the operators are trained to work on trees.

- **Differential evolution (DE)**: The DE algorithm is simpler and easier to use than GA. It generates solutions by relying on differences between mutation and crossover. Its core is to compute differences between two random chromosomes from the population.
- **Evolution strategy (ES)**: The solutions proposed by ES are based on the suitability of chromosomes. It differs from GA concerning the selection procedure. Unlike GA, which relies on a parental population, ES creates a new and temporary population where each chromosome undergoes crossover and mutation.
- **Evolutionary programming (EP)**: Originally discovered to evaluate and improve grammar in unknown languages, EP is now more popular as a tool for optimising numerical problems.

3 Data-Driven Modelling and Simulators in Polymer Science

The application of AI in material science is dynamic. Incorporating ML has enabled rapid discovery, design and optimisation of intended products. Its revolutionary assistance is not only defined by its ability to process large datasets but also by its capacity to integrate data-driven modelling with simulation for materials. This approach through modern-day AI has enabled rapid innovation in material designing for optimization and quality control of sustainable polymer production.

(1) **Data collection**: Data collection for ML model is the prime and crucial step for successful material designing. The amount of data collected determines the efficiency and robustness of the study. Traditionally, data reported in published research articles were preferred because they have higher reliability. However, with modern advancement, extensive databases are available to retrieve polymer properties. Polymer data is either numerical (e.g. name and property value) or structural. Numerical data is usually stored in Excel files or other tabular form for easier access and utilization. Alternatively, structural data possess specific traits and is represented through file formats. Some commonly used extensions for polymer structural data are listed below [8–10].

   - **'.mol'**/ **'.sdf'** (MDL molfile/structural-data file formats) are widely used for the storage of molecular structures, like atom coordinates, bond information and other related properties.
   - **'.pdb'** (protein data bank) is mainly used to represent 3D structures of biomolecules (proteins, nucleic acids, etc.), related properties and experimental data.
   - **'.smiles'**/ **'.smi'**(Simplified Molecular Input Line Entry System/SMILES) presents structures of molecules in line notation. It is simplified for human interpretation and enable easy processing of data.
   - **'.xyz'** refers to molecular structures in XYZ format. It contains atom coordinates and is easily processed by visualisation software.

(2) **Polymer descriptors**: The data is quantified and represented as a numerical. These are readable by computer and are commonly known as **polymer descriptors**. There are thousands of such descriptors that contain essential information about polymer structures. These descriptors are ensured to hold as accurate information as possible because they hold input values offered to ML models. Hence, the success of polymer design relies heavily on the descriptors. Since there is no optimal method for the determination of descriptors, they are commonly selected through comparison by closely associating them with target properties [8, 9].

Despite being in large amounts, polymer descriptors can be classified into **monomer-level descriptors**, which focuses on features of monomers, like molecular weight, carbon backbone, chemical composition, structural properties and functional groups; and **bulk material descriptors**, which highlights larger features, such as physiochemical properties, chian length and surface features.

The polymer descriptors can be computationally generated using softwares like **Dragon** (largest in variety with over 5000 2D and 3D descriptors) and **RDkit** (Python package for molecular representation and calculation; over 200 descriptors).

(3) **ML approaches**: Quantitative structure–property relationship (QSPR) of polymers is obtained by utilising different ML algorithms like decision trees, SVM, random forest, ANN, GA, etc.
(4) **Polymer development**: The polymer descriptors and target property values are supplied to the ML model as input data and output data, respectively. The model proceeds to generate as accurate prediction as possible by filtering polymer candidates based on the target. Finally, a suitable polymer with target characteristics is identified for laboratory synthesis.

Traditional methods of polymer designing have relied heavily on experimentation on physics-based models to direct the production of materials. Researchers would synthesise polymers with varied formulations to analyse based on trial-and-error. Additionally, classic computational simulations have supported in modelling polymer properties and behaviour under various conditions for a long time.

A simulation technique called finite element analysis, or FE is used to forecast how a product would behave in real-world situations. It is based on the finite element technique (FEM) and can be used to analyze the product's structure's dynamics, durability, and heat distribution. Similarly, using a broad model of interatomic interactions, molecule dynamics (MD) simulated the motion of atoms and molecules and forecasted their future motion based on that study. These approaches, however, have been over simplified and frequently predicated on presumptions that fail to adequately account for the multifaceted nature of polymers. These heuristic approaches needed a lot of resources, which resulted in expanded period of development even if they generated significant discoveries and advancements. However, these methods are simplistic and often based on assumptions that does not fully capture the complexity of polymers. Despite delivering substantial advancements and innovations, these heuristic approaches required extensive resources, which led to prolonged development over long periods of time. These did not succeed in capturing the complex modelling of polymers and often fell short during industrial-scale productions, causing substantial loss of resources. Conversely, AI-based methodologies have led to a shift towards efficiency and accuracy. It has dramatically reduced the time required for screening and optimisation. The level of precision for creating polymers with complex properties has significantly increased while being economically efficient by minimising the overall production cost. Data-driven methods help us monitor production lines in real-time and find defects quickly. However, it's important to continuously use high-quality data for accurate simulations and predictions [10–13].

Novel AI-based methodologies for polymer and plastic designing rely extensively on assessing the vast number of datasets, consisting of material properties, historical experimental results, and data collected through real-time analysis of different polymers (Table 1). Various machine learning algorithms are modelled within complex sensor networks to utilise these datasets during processing, performance testing, and quality control in material synthesis.

Table 1 ML models and databases which specialise and aid in polymer designing

Database/ML model	Description
PolyID	It is an algorithm for identifying materials accessible from waste substrates to design biobased polymers. It navigates numerous renewable polymers to develop enhanced and sustainable materials. It was developed by coupling neural networks with polymer structure representation
RadonPy	It is a Python-based (Python 3.7–3.9) package to be used jointly with the Python library RDkit. It is designed specifically for supercomputers where multiple polymers can be independently calculated in parallel. It automates processes to perform classical molecular dynamics simulations
PolyInfo	It is the largest polymer database, with more than 20,000 polymers and over 100 polymer properties. It stores the chemical structures of different polymers, their IUPAC names, properties under measurement conditions, processing methods, and more
PI1M	It is a hypothetical database curated from PolyInfo where approximately 12,000 polymers were collected manually and modelled to generate a dataset with around a million polymers
CRIPT	Community Resource for Innovation in Polymer Technology This is a data structure that supports the generation of metadata from experimentation and system simulations. It is modelled to scale big data while maintaining efficiency in searching and saves storage by lowering data repetition
PolyDAT	It is a generic schema for the structural characterisation of polymers. It is designed to contain multi-species reaction network construct to create most comprehensive profile for a particular polymer species. It serves to provide and share necessary quantitative information about polymer characterisation among researchers
PolyG2G	It is designed to generate graph-to-graph model specifically for polymers. It utilises graph neural networks for creating new polymers structures based on learned patterns

Modelled polymers are simulated using data collected from real-time sensors to allow dynamic adjustments before processing. This ensures consistency and control over the intended product properties while lowering the chances of the defect. For example, GA optimised challenges related to solar array layouts, which enhanced power output and increased installation flexibility. GP has been utilised to predict pavement deterioration. It was trained with a dataset generated from real-world data from the roads of Patiala (Punjab, India). It used five models to predict the progression of distress, such as roughness, ravelling, cracking, rutting, and potholes. The intelligent battery proposed by Fadda et al. exploits a customised ES algorithm to adapt its recharging strategy to varying environmental conditions.

4 Case Studies and Practical Applications

Case Study 1: Machine Learning and Genetic Algorithm for designing recyclable plastics.

Plastics are an integral and problematic part of modern society. Despite being a convenient and cost-effective packaging solution, they are a huge environmental threat. These synthetic materials take extraordinary time to decompose and exist for hundreds of years. Microplastics released as a result of the breakdown of bigger plastics enter the food chain and accumulate in animals and humans, leading to various diseases. This study intended to develop an alternative to polystyrene, one of the most widely use plastics. This choice also arose from the toxicity concerns contributed by the monomer, styrene.

A novel GA was tailored to design chemically recyclable polymers. Hypothetical monomers were designed and subjected to virtual templates for generating polymers. GA was implemented in following steps:

- Generation of scaffold which comprised of various functional sites and can act as a base structure capable of polymerisation
- A list of the global R-groups is created
- These R-groups are visualised to bond with the generated scaffold to created polymerizable molecules
- Smiles Arbitrary Target Specification (**SMARTS**) is used to polymerise the formerly generated molecules using VFS virtually.

The initiation is followed by the creation of *n* molecules from previous scaffolds using GA and assigning them to random R-groups to form functional molecules. Each molecule undergoes polymerisation, and its properties are predicted using another ML model. All the generated polymers are then ranked based on fitness, and the top-ranking polymer is labelled as "parents" for succeeding generations. Each parent polymers are combined randomly to generate a finite number of children. Since each child inherits one of the parent scaffolds, they are further crossed over to combine the parent's R-groups. As a result, some functionalisation sites are replaced by random R-groups and a small subset of children are mutated. This mutation persists until a unique molecule is generated and fingerprinted. The most desirable polymer of that generation becomes "parents". This process is repeated to drive the generation of new and evolved polymers.

This study introduced a GA specifically designed to create monomers that can be chemically recycled and act as an alternative to polystyrene. This GA was combined with BRICS-derived fragments and VFS techniques to pinpoint 75 polymers that could act as potential alternatives for polystyrene. They exhibited the required mechanical, thermal and thermodynamic properties necessary for it to be recyclable. Although it served exceptionally as a prediction tool, it failed to suggest a synthesis pathway for the generation of the modelled monomers.

Case Study 2: Discovery of high-temperature dielectrics for energy storage through AI.

This study utilised AI and discovered new dielectrics in polyimide and poly-norbornene families. **PolyVERSE** (**Poly**mers designed by**V**irtually-**E**xecuted**R**ule-Based**S**ynthesis**E**xperiments) was relied on to succeed in attaining four target attributes:

An expansive subspace, enough to unravel unknown materials
Limit false positives, labelled as challenging to synthesise hypothetical materials
Accurate property estimation
Efficient prediction

The design of the model was approached in salient steps.

a pool of various chemical structures was generated, and their properties were predicted. The best candidates were screened and synthesised for characterisation.
Chemical structures were generated in three steps: curating a database for available monomers, choosing a reaction to select or reject each monomer, and finally, chemically transforming selected monomers into polymers.
The structures are screened on the basis of carefully chosen and application-specific filters, which are high glass-transition temperature, band gap and dielectric constant.

Regrettably, retrieving a product with all three properties simultaneously was not attained. But PONB-2Me5C1 was discovered, which surpassed current commercial dielectrics, particularly at elevated temperatures.

Case Study 3: AI-aided structure optimisation for enhanced gene delivery.

This study used poly(β-amino ester) or PAE as a polymerization vector as a case study to examine how polymer structure affects gene delivery. The structure and functions of polymer components employed as vectors in gene therapy have been revealed together with the corresponding impacts on gene transport. To decipher the connection between transfection of genes efficacy and polymer component distribution (PCD), artificial intelligence (AI) analysis was utilized. Through AI analysis, the effects of different polymer constituents and their distribution on gene delivery efficiency were evaluated.

5 AI for Drug-Discovery

The market for AI polymers has been significantly affected by COVID-19. The pandemic hastened the drug industry's use of machine learning. Several of these algorithms efficiently analyze throughout massive datasets to create medications by forecasting their molecular makeup, activity, and possible adverse effects. In drug delivery systems, polymers such as the hydrogels, implants, especially encapsulation vehicles are important. Polymer-based carriers, which have improved biocompatibility and regulated release for better therapeutic response, are being designed and launched using AI-driven models [6–9].

6 Challenges and Future Perspectives

In order to fully employ AI in polymer design, a few issues need to be resolved despite the significant progress made in polymer informatics. It covers everything from model development, technical limitations, and data accessibility to quality and regulatory issues.A major challenge is the need for an extensive, standardised database of polymer properties and most of the available datasets are scattered, based on diverse experimental setups and reported in different formats. The lack of integrated standards for reporting polymer properties effects the development of accessible and efficient models. Developing and maintaining AI models comes with a fair share of risks and challenges. Transitioning polymer products AI-modelled in the laboratory to industrial-scale manufacturing requires substantial capital investments and a precise operational system. Despite being less prone to defects and errors, AI models are vulnerable to data poisoning, resulting in a heavy loss of resources if not detected on time. The products of Edisonian methods are diverse, and each predicted model undergoes a separate experimental route, making them prone to errors and easier to diagnose for defects simultaneously. High efficiency and performance necessitate training complicated algorithms on big datasets. Smaller businesses or institutions cannot afford this because it requirements a significant financial outlay. On the contrary hand, creating a model with fewer variables will lead to substandard algorithms that will undoubtedly not be able to survive weaknesses and produce subpar outcomes.

With the goal to encourage the responsible development and widespread implementation of AI throughout the European Union, the EU Artificial Intelligence Act may come into effectiveness based on demands from users. It established a uniform framework based on the "risk" approach, which emphasizes minimal risk (allowing non-obligatory companies to implement additional codes of conduct), specific transparency risk (requiring chatbots to clearly inform consumers of the interacting machine), high-risk (requiring systems, like AI-based medical software, to adhere to stringent requirements), and unacceptable risk (prohibiting systems that clearly violate fundamental rights). This ensures an improved AI ecosystem for services that are safer and better performing. In a similar vein, more specific requirements for rights in intellectual property are required for determining ownership and patents for different polymer inventions.

AI has made the process simpler to categorize and understand numerous kinds of plastic. This leads to less pollution and higher processing efficiency. The algorithms improve the ability to detect of plastic through using infrared spectroscopy and sophisticated optical sensors. By precisely identifying and controlling the conditions required for maximizing both the yield and the standard of the products, they additionally strengthen thermochemical conversion techniques.

7 Future Directions

The demand for AI-driven polymers is driven by their effectiveness and cost-effective solutions in drug discovery and materials science. This often results in discussions for its improvement. Various literatures have discussed and hypothesised hybrid algorithms by merging models. These novel algorithms will be modelled by incorporating complex networks of databases and design frameworks to improve predictive accuracy. For example, memetic algorithms stand out as one such interesting area for future exploration.

Through innovative polymer growth, recent developments in computer science and other fields are changing our ability to address health and environmental concerns. By guiding medications to particular parts of the body, nanoparticles, for illustration, have shown great promise in targeted drug delivery, enabling precise treatments. Preventing these particles' bioaccumulation and biomagnification after they transported the chemical is still an important challenge, though. For the next generation of treatment choices, it is essential to develop nanotechnology that not merely carry medication but also safely decay or disperse without triggering adverse consequences. Similarly, there is a vital requirements for sustainable, biodegradable substitutes for conventional plastics. AI-driven methods are rapidly advancing the identification of eco-friendly polymers that offer high performance and degradation. By using advanced computational models and data-driven simulations, researchers may enhance material properties with accuracy to attain desired properties or characteristics. These advancements pave the way for next-generation polymers that help attain customised, targeted, eco-friendly and sustainable polymer products.

8 Inference

An important development in materials science is the integration of AI into polymer design. Traditional methods' accuracy and efficiency are significantly increased by the answers it offers. AI-driven techniques, which make use of sophisticated computational models and large datasets, have sped up the identification of polymers and greatly improved their characteristics, resulting in discoveries such as biodegradable plastic and effective dielectrics. The lack of a robust and organized regulatory body and norms, substantial technical needs, and a lack of adequate data sources are still obstacles.

Future studies are reportedly concentrating on hybrid algorithms that augment predictive capacity by fusing artificial intelligence (AI) with traditional simulation techniques, creating sophisticated medication delivery systems, and substituting polymers with environmentally friendly materials.

Incorporating AI into polymer design signifies a major transformation in materials science. It provides solutions that greatly improve the efficiency and precision of traditional techniques. By utilizing advanced computational models and extensive datasets, AI-driven methods have accelerated the discovery of polymers and significantly enhanced their properties, leading to innovations with like efficient dielectrics and biodegradable plastic. Nevertheless, challenges like insufficient data sources, significant technical requirements, and the absence of a strong and structured regulatory body and guidelines remain.

Future researches are observed to be focused on hybrid algorithms that combine AI with classical simulation methodologies to enhance predictive ability, producing advanced drug delivery systems and replacing polymers with poor environmental impact. Ultimately, the ongoing development of AI in polymer design promises to improve product performance and manufacturing efficiency and support sustainability, thus shaping the future of polymer informatics and wider materials science.

Acknowledgements The authors are thankful to the Director, CSIR-IITR, Lucknow, for their continuous encouragement and access to a rich knowledge base at the Knowledge Resource Centre (KRC).

Conflict of Interest The authors declare no conflict of interest.

References

1. Atasi, C., Kern, J., Ramprasad, R.: Design of recyclable plastics with machine learning and genetic algorithm. J. Chem. Inf. Model. **64**(24), 9249–9259 (2024). https://doi.org/10.1021/acs.jcim.4c01530
2. Bejagam, K.K., Gupta, N.S., Lee, K.-S., Iverson, C.N., Marrone, B.L., Pilania, G.: Predicting the mechanical response of polyhydroxyalkanoate biopolymers using molecular dynamics simulations. Polymers **14**(2), 345 (2022). https://doi.org/10.3390/polym14020345
3. China C.R.: Five Machine Learning Types to Know. https://www.ibm.com/think/topics/machine-learning-types (2023)
4. Chopra, T., Parida, M., Kwatra, N., Chopra, P.: Development of pavement distress deterioration prediction models for urban road network using genetic programming. Adv. Civil Eng. **2018**(1), 1253108 (2018)
5. Convolutional Neural Networks (CNNs), Deep Learning, and Computer Vision. https://www.intel.com/content/www/us/en/internet-of-things/computer-vision/convolutional-neural-networks.html

6. Fadda, E., Perboli, G., Squillero, G.: Adaptive batteries exploiting on-line steady-state evolution strategy. Proc. Europ. Conf. Appl. Evolut. Comput. LNCS **10199**, 329–341 (2017)
7. Georgia Tech University, Atlanta, GA.: Using Artificial Intelligence to Find the Polymers of the Future. https://www.mobilityengineeringtech.com/component/content/article/51780-adttst-10241#:~:text=While%20AI%20can%20accelerate%20the,to%20rapidly%20explore%20new%20ideas (2024)
8. Hayashi, Y., Shiomi, J., Morikawa, J., et al.: RadonPy: automated physical property calculation using all-atom classical molecular dynamics simulations for polymer informatics. npj Comput Mater **8**, 222 (2022). https://doi.org/10.1038/s41524-022-00906-4
9. Lin, T.S., Rebello, N.J., Beech, H.K., Wang, Z., El-Zaatari, B., Lundberg, D.J., Johnson, J.A., Kalow, J.A., Craig, S.L., Olsen, B.D.: PolyDAT: a generic data schema for polymer characterization. J. Chem. Inf. Model. **61**(3), 1150–1163 (2021). https://doi.org/10.1021/acs.jcim.1c00028
10. Lv, M., Li, J., Du, H., Zhu, W., Meng, J.: Solar array layout optimization for stratospheric airships using numerical method. Energy Convers. Manage. **135**, 160–169 (2017)
11. Slowik, A., Kwasnicka, H.: Evolutionary algorithms and their applications to engineering problems. Neural Comput. Appl. **32**, 12363–12379 (2020). https://doi.org/10.1007/s00521-020-04832-8
12. Stryker, C., Kavlakoglu, E.: (2024). What is Artificial Intelligence (AI)?. https://www.ibm.com/think/topics/artificial-intelligence (2024)
13. The Impact of AI on Polymer Research https://growthgarage.mcgc.com/resources/the-impact-of-ai-on-polymer-research#:~:text=AI%20can%20assist%20in%20accelerating,strength%2C%20flexibility%2C%20or%20biocompatibility
14. Walsh, D.J., Zou, W., Schneider, L., Mello, R., Deagen, M.E., Mysona, J., Lin, T.S., de Pablo, J.J., Jensen, K.F., Audus, D.J., Olsen, B.D.: Community resource for innovation in polymer technology (CRIPT): a scalable polymer material data structure. ACS Cent. Sci. **9**(3), 330–338 (2023). https://doi.org/10.1021/acscentsci.3c00011
15. Wilson, A.N., St John, P.C., Marin, D.H., Hoyt, C.B., Rognerud, E.G., Nimlos, M.R., Cywar, R.M., Rorrer, N.A., Shebek, K.M., Broadbelt, L.J., Beckham, G.T., Crowley, M.F.: PolyID: artificial intelligence for discovering performance-advantaged and sustainable polymers. Macromolecules **56**(21), 8547–8557 (2023). https://doi.org/10.1021/acs.macromol.3c00994
16. Zhao, Y., et al.: A review on the application of molecular descriptors and machine learning in polymer design. Polym. Chem. **14**(29), 3325–3346 (2023). https://doi.org/10.1039/d3py00395g
17. Gurnani, R., Shukla, S., Kamal, D., et al.: AI-assisted discovery of high-temperature dielectrics for energy storage. Nat. Commun. **15**, 6107 (2024). https://doi.org/10.1038/s41467-024-50413-x
18. Uddin, S., Haque, I., Lu, H., et al.: Comparative performance analysis of K-nearest neighbour (KNN) algorithm and its different variants for disease prediction. Sci. Rep. **12**, 6256 (2022). https://doi.org/10.1038/s41598-022-10358-x
19. Regulation (EU) 2024/1689 of the European Parliament and of the Council of 13 June 2024 laying down harmonised rules on artificial intelligence and amending Regulations (EC) No 300/2008, (EU) No 167/2013, (EU) No 168/2013, (EU) 2018/858, (EU) 2018/1139 and (EU) 2019/2144 and Directives 2014/90/EU, (EU) 2016/797 and (EU) 2020/1828 (Artificial Intelligence Act) (Text with EEA relevance) https://eur-lex.europa.eu/eli/reg/2024/1689/oj/eng
20. Li, Y., He, Z., SA, Wang, X., Li, Z., Johnson, M., Foley, R., Sáez, I.L., Lyu, J., Wang, W.: Artificial Intelligence (AI)-aided structure optimization for enhanced gene delivery: the effect of the polymer component distribution (PCD). ACS Appl. Mater. Interf. **15**(30), 36667–36675 (2023). https://doi.org/10.1021/acsami.3c05010

Structure–Property Relationships in Polymers

Michael Ovbare Akharame, Aimuanmwosa Frank Eghomwanre, Fidelis Ifeakachukwu Okolafor, Bamidele Oladapo Fagbayigbo, and Ogheneochuko Utieyin Oputu

Abstract Polymers are a diverse and versatile class of materials composed of long-chain macromolecules made up of repeating structural units known as monomers. These materials display a vast array of properties that range from flexible and elastic to rigid and glassy, largely due to their structural variability across different length scales. The term structure–property relationship in polymer science refers to the correlation between the molecular, supramolecular, and macroscopic structural characteristics of a polymer and its physical, mechanical, thermal, chemical, optical, and barrier properties. The study of structure–property relationships in polymers has been a cornerstone of polymer science, driving innovation across a wide array of applications, from biomedicine to electronics and sustainable materials. By understanding the underlying principles of how molecular and macromolecular structures influence material properties, researchers can design polymers with tailored characteristics that meet the increasingly complex demands of modern industries. This chapter focused on the importance of these relationships in diverse polymer systems, X-raying in detail the molecular structures and types of polymers, the key properties of polymers, and the structure–property relationships in polymeric materials. Moreover, the challenges and future directions in structure–property engineering are discussed.

M. O. Akharame (✉) · A. F. Eghomwanre
Department of Environmental Management and Toxicology, University of Benin, Benin-City, Nigeria
e-mail: michael.akharame@uniben.edu

A. F. Eghomwanre
e-mail: frank.eghomwanre@uniben.edu

F. I. Okolafor
Department of Science Laboratory Technology, University of Benin, Benin City, Nigeria
e-mail: fidelis.okolafor@uniben.edu

B. O. Fagbayigbo
Department of Environmental and Occupational Health, Cape Peninsula University of Technology, Cape Town, South Africa

O. U. Oputu
The Hair and Skin Research Laboratory, Division of Dermatology, Department of Medicine, University of Cape Town, Cape Town, South Africa

B. Sharma et al. (eds.), *AI-Driven Polymer Design*, Engineering Materials,
https://doi.org/10.1007/978-981-95-8353-9_2

Keywords Polymers · Polymeric materials · Polymer structure · Polymer structure–property relationships

1 Introduction

Polymers and polymeric materials are increasingly being utilized globally in virtually all aspects of human life. These include utilization as packaging and wrapping materials, clothing manufacturing, construction materials and accessories, medical and healthcare supplies, electronics and electrical components, and automobile and aeronautic parts. The numerous polymers and polymeric composites possess unique and intrinsic properties that make them suitable for their designated area of application [1]. The properties considered in polymers for specific purposes may include their electrical, mechanical, thermal, optical, and magnetic properties. In certain instances, the properties of interest may be gas and solvent transport, chemical stability, hydrophobic/hydrophilic balance, or optoelectronic properties [2].

Monomers are the building blocks of polymeric structures; they form a network of repeat units, producing large molecules or macromolecules [3]. The characteristics and properties of a polymer are largely determined by the monomer units; factors such as the nature, amount, and arrangement of the monomer are critical determinants [4]. Polymers are essentially either of natural or synthetic origin. These two categories broadly cover the numerous polymer resources available to humans. Polymers of natural origin have existed and were utilized by early humans in basic ways. By the nineteenth century, a period of major industrial revolution, chemical processing and enhancement of natural polymers were initiated for better outputs and usage compatibility. Vulcanization is one such process that makes natural rubber more useful. Synthetic polymers are among the innovative and developmental milestones achieved in recent decades (twentieth century). Synthetic polymers are chemically synthesized from monomer units via a process called polymerization. Synthetic polymer creation through chemical reactions provides boundless opportunities for synthesizing different polymer matrices or blends. Chemical processes offer room for matrix modification and functionalization to suit expected usage requirements and conditions. Synthetic polymers have various advantages, such as low synthesis costs, enhanced thermomechanical properties, and high chemical resistance [4].

The vast and continuous use of polymers in today's world is hinged on the development and innovative drive with the ability to synthesize matrices that meet specific application conditions and demands. The structure–property relationships in polymers play an important role in ensuring that designed polymers meet specific functionalities. All polymers, whether natural or synthetic, consist of a backbone of multiple covalently bonded atoms; their differences emanate from the monomer type, bond structures and networks, as well as the relative strength of the intermolecular secondary valence bonds [5]. The monomer configuration and conformation along the backbone contribute significantly to the final properties obtained, especially for copolymers [6]. Essentially, polymers' inherent properties, which range from flexible and elastic to rigid and glassy, are largely influenced by their structural variability across different length scales. Consequently, this chapter focuses in detail on the structure–property relationships in polymeric materials, which play crucial roles in the key properties of polymers. Correlations exist between the molecular, supramolecular, and macroscopic structural characteristics of a polymer and its physical, mechanical, thermal, chemical, optical, and barrier properties.

2 Molecular Structures and Types of Polymers

Polymers are high-molecular-weight compounds made up of repeating smaller units known as monomers. These monomers are chemically bonded through covalent linkages to form long, chain-like macromolecules that can vary in complexity and structure. The nature of these chains, including their length, branching, and arrangement, plays a significant role in determining the physical and chemical properties of the resulting material. Polymers exist in both natural and synthetic forms. Natural polymers include substances such as cellulose, which is found in plant cell walls, and proteins, which are essential biological macromolecules composed of amino acid chains. Synthetic polymers, on the other hand, are human-made and engineered for specific purposes [7]. Common examples include polyethylene, which is widely used in plastic bags and packaging, and polystyrene, which is used in disposable containers and insulation materials. These synthetic polymers have become vital materials in various industries due to their versatility, durability, and cost-effectiveness [8].

3 Types of Polymers

Polymers are essentially derived from either natural (made from plant and animal tissues) or synthetic (made from fossil resources) sources. The demand for both natural and synthetic polymers in a variety of sectors, such as the packaging industry, is high due to their inherent properties [9].

3.1 Natural Polymers

Natural polymers are sourced from organic substances, and they are generally acceptable and widely utilised in several applications. Natural polymers have been utilized for 1000 years but initially only in very small quantities, such as when plant gums were employed as wood glue for houses or as a waterproof substance for boats [10]. Natural polymers, such as chitosan/chitin, starches, cassava, maize, gums, pectin, wood, potatoes, straw, and wheat, are found in both plant and animal remains and

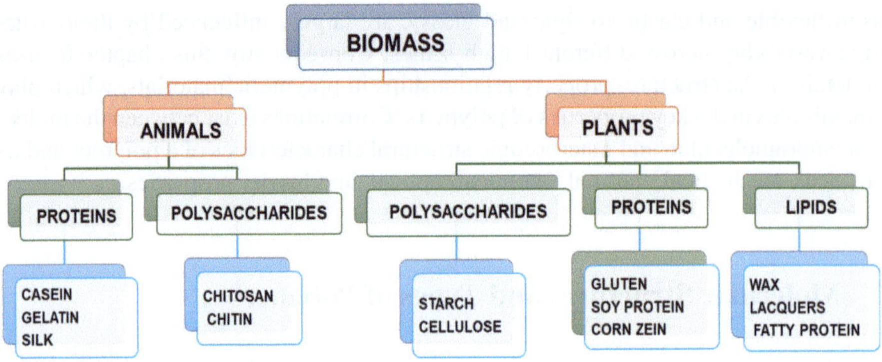

Fig. 1 Hierarchical distribution of natural polymers derived from biomass

are used to make these biodegradable polymers [11]. Polymers contain polysaccharides and are found in nature as microbes (e.g., xanthan gum, dextran), animals (e.g., chitosan, chondroitin), plants (e.g., guar gum, pectin) and algae (e.g., alginates). Natural polymers are environmentally friendly polymeric materials with renewability capacity. Proteins and nucleic acids are the major constituents of natural polymers [12]. The protein constituents of natural polymers perform several functional and structural roles that are essentially necessary for life, whereas nucleic acids mostly serve as natural storage and transmission of genetic information from one generation to another in terms of inheritance [13]. Natural polymers are essential for life and play crucial roles in biological processes. The hierarchical distribution of natural polymers derived from biomass is presented in Fig. 1.

3.2 *Synthetic Polymers*

Synthetic polymers are man-made polymeric materials primarily sourced from petroleum chemicals in a controlled setting. They are materials made of high-molecular-weight substances composed of repeating units of monomers joined through chemical reactions [14]. Synthetic polymers are specially engineered to meet the specific needs of various industries, such as the application packaging industry (plastic), textile industry (clothing and dyes), health industry (medicines and surgical materials), and technology industry (electronics materials), among other applications [13]. Synthetic polymers are essentially divided into three classes, namely, thermoplastics, thermosetting plastics, and elastomers (Fig. 2), depending on how they are structurally and chemically bonded. Thermoplastic polymers include some of the most important groups of plastics, such as polyethylene, polypropylene, and polystyrene polyvinyl chloride, among others [15]. Thermoplastics are polymer materials that melt when exposed to increased temperatures and can be

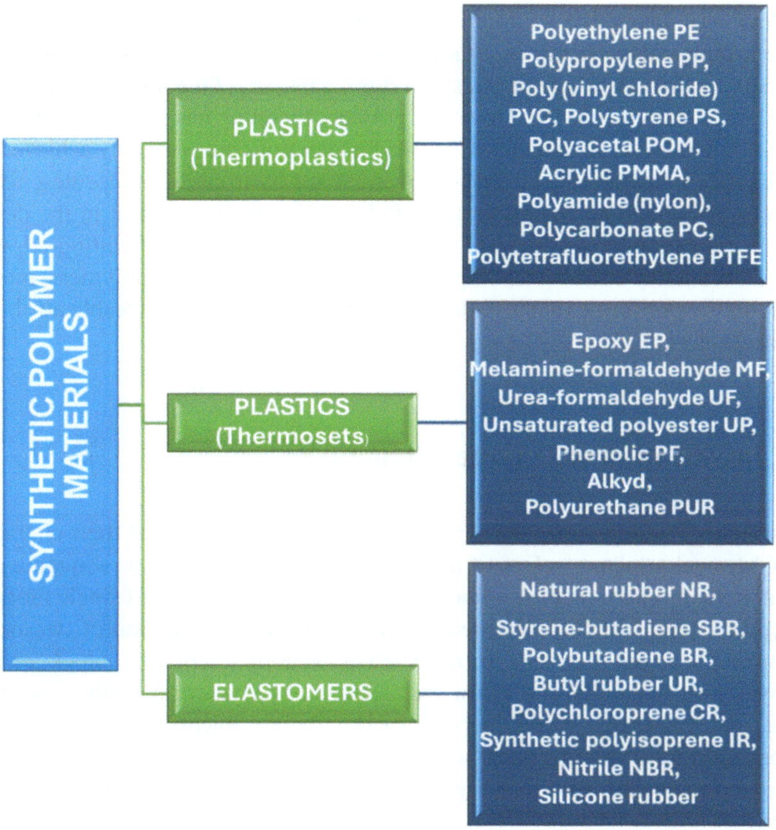

Fig. 2 Classification of synthetic polymers

processed by a variety of moulding and extrusion techniques, whereas thermosetting polymers cannot be melted or re-moulded. Examples of thermosetting polymers include alkyds, amino and phenolic resins, epoxies, polyurethanes, and unsaturated polyesters. Elastomers are polymers with inherent high elasticity due to their coiled structure; they are rubber-like solids with elastic properties. Synthetic polymers offer low-cost packaging designs for easy transportation of foods, medicine, and goods such as chemicals, cosmetics, detergents, and other packaging needs [13].

3.3 *Polymer Blending*

In an attempt to improve the quality of polymeric materials with limited economic value, the physicochemical characteristics of both natural and synthetic polymers have been exploited to develop unique polymeric materials [16, 17]. Polymer blending was developed to improve synthetic polymers with respect to the need for environmental implications. Studies have revealed that the production of blended polymers, such as short-life plastics, is a significant part of the mitigation of environmental implications [18, 19]. However, polymer blending has several shortcomings, despite its numerous advantages. Chemical interactions are not easily achieved since the majority of naturally occurring polymers are hydrophilic molecules, whereas synthetic polymers are lipophilic [20]. Polymer blending is useful in the development of unique materials, modern manufacturing processes, and cutting-edge technology. Studies have shown the importance of blending different polymeric materials at specific proportions to achieve desirable qualities to create noticeable attributes of new materials [17, 21].

4 Classification of Polymer Structures

The structural features of polymers are classified on the basis of their characteristics. Chemical structure refers to the type of monomer(s) and the sequence in the polymer chain. These include homopolymers formed as a result of single monomers and copolymers, which are formed as a result of the combination of two or more monomers arranged in diverse forms, such as random, alternating, block, and graft copolymers. Essentially, the chain architecture of polymers describes the diverse shapes and branching of polymers. For example, the linear form of a polymer is composed of long straight branches with no branches [22]. The polymerization of macromonomers also results in regular multibranched polymers with a high branch density.

Polymer structures vary widely, and their architecture significantly influences their physical and mechanical behaviour;

- **Linear polymers** consist of monomer units joined end-to-end in single, continuous chains. These structures, such as high-density polyethylene (HDPE), typically allow for close packing and crystallinity, resulting in increased strength and melting points [23].
- **Branched polymers** have side chains attached to the main backbone. This irregular structure reduces chain packing and crystallinity, often resulting in materials that are more flexible but less dense, such as low-density polyethylene (LDPE) [8].
- **Crosslinked polymers** contain covalent bonds between polymer chains, forming a three-dimensional network. This structure increases elasticity, thermal stability, and resistance to solvents, as observed in vulcanized rubber [24].

Additionally, the type of monomer and the polymerization method used significantly influence the resulting polymer's structure and properties. For instance, the presence of functional groups in monomers can introduce polarity, reactivity, or hydrogen bonding, which affects thermal behavior and solubility [24]. Polymerization typically occurs through two main mechanisms: addition polymerization (also

Table 1 Summary of the polymer structure, bonding, thermal behaviour, density, and properties

Polymer structure	Bonding	Thermal behaviour	Density	Properties
Linear	Weak intermolecular (van der Waals, H-bonds)	Thermoplastic (softens on heating)	High-density, efficient packing	Flexible, less rigid
Branched	Weak intermolecular; branches do not covalently bond to other chains	Mostly thermoplastic; some complex branches are thermosetting	Lower density due to branching	More flexible, less dense
Crosslinked	Covalent bonds between chains (crosslinks)	Thermosetting (do not melt)	Moderate to high-density	Rigid, strong, thermally stable
Networked	Extensive covalent 3D network	Thermosetting (do not soften)	Very high density, rigid	Very rigid, often brittle

known as chain growth) and condensation polymerization (or step growth). In addition polymerization, it involves the use of monomers with double bonds or reactive groups to form polymers without the loss of small molecules, such as polyethylene and polystyrene, are classic examples. In contrast, condensation polymerization involves monomers reacting with the elimination of small byproducts, such as water or methanol, as observed in the formation of polyesters and polyamides [8]. Importantly, the stereochemistry of a polymer, which represents the spatial arrangement of side groups attached to the polymer backbone, reflects the degree of crystallinity and mechanical characteristics [25]. A summary of the polymer structure, bonding, thermal behaviour, density, and properties is listed in Table 1.

5 Key Properties of the Polymers

The diverse functionalities of polymers arise from the rich interplay between their structural characteristics and resulting material properties. Understanding these relationships is essential for designing polymers with targeted performance characteristics across various applications, from biomedical devices to aerospace composites. Some common polymers and their key properties are presented in Table 2.

Table 2 Common polymers and their key properties

Polymer matrix	Chemical formula	Structure of monomer unit	Key properties of polymers
[b]Polyamide	$(CONH_2)n$	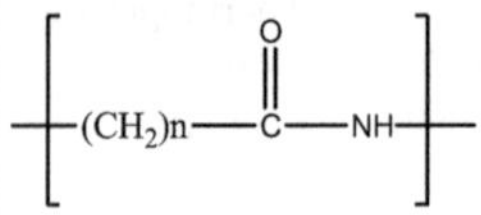	• High tensile strength and toughness • Excellent resistance to wear and abrasion • Resistance to a wide range of chemicals, oils, and solvents • Low water absorption rate • Good insulation properties
[a]Polyimide	$(CO\text{-}NR_2)n$	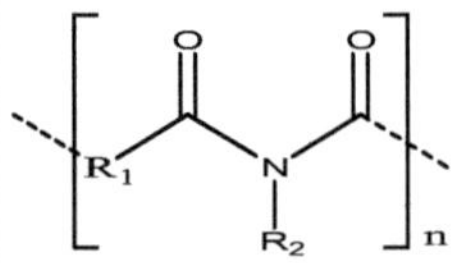	• High heat resistance and thermal conductivity • High tensile strength, flexibility, and toughness • Excellent chemical resistance • Good electrical insulation • Resistance to radiation
[b]Polystyrene	$(C_8H_8)n$	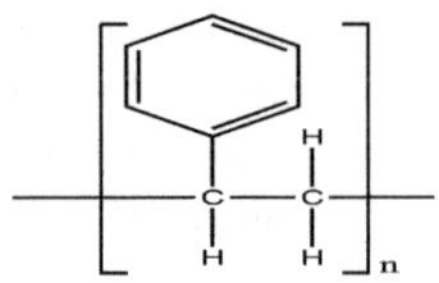	• Lightweight, stiff, rigid, and transparent • High tensile and flexural strength • Low impact strength • Low heat resistance • Good electrical insulation
[b]Polyacrylamide	$(C_3H_5NO)n$	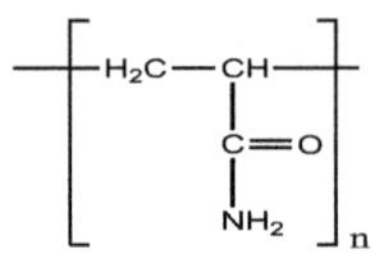	• High water absorption and hydrophilicity • Good adhesiveness and non-toxicity • Moderately stable to heat • Possesses the ability to form hydrogels

(continued)

Table 2 (continued)

Polymer matrix	Chemical formula	Structure of monomer unit	Key properties of polymers
[b]Polysulfone	$(C_{27}H_{26}O_6S)n$	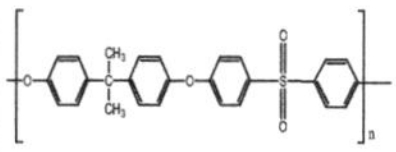	• High heat resistance and low thermal expansion • Excellent heat deflection temperature • High strength and stiffness, good impact resistance, and transparent • Excellent creep resistance • Resistant to acids, bases, solvents, oxidising agents • Good radiation resistance and adhesion to metals
[b]Polypyrrole	$(C_4H_5N)n$	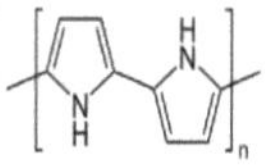	• High electrical conductivity • Good mechanical stability but exhibits brittleness • Good redox properties • Good stability in its oxidised form • Exhibits good biocompatibility
[b]Polydimethyl siloxane	$(C_2H_6OSi)n$	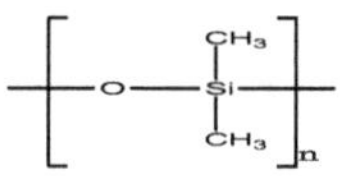	• High flexibility and elasticity • Exhibits good biocompatibility and low toxicity • High chemical inertness and resistance to degradation • Possesses optical transparency and low autofluorescence • High gas permeability
[b]Polyvinyl alcohol	$(C_2H_4O)n$	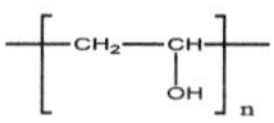	• Good mechanical strength and durability • High water solubility and hydrophilicity • Resistant to oils, greases, and solvents • Possesses film-forming ability and effective oxygen and aroma barrier • Exhibits good biocompatibility and non-toxicity

(continued)

Table 2 (continued)

Polymer matrix	Chemical formula	Structure of monomer unit	Key properties of polymers
[b]Polyacrylonitrile	$(C_3H_3N)n$	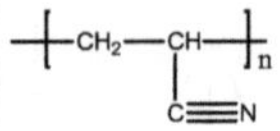	• High tensile strength and tensile modulus • High thermal stability • Inert and resistant to most organic solvents and acids • Possesses light and UV resistance • Possesses good thermal stability
[b]Polyethersulfone	$(C_{12}H_8OSO_2)n$	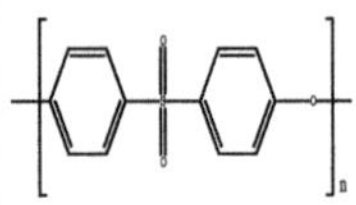	• Excellent mechanical strength, toughness, and fatigue resistance • Possesses optical transparency and is resistant to many acids, alkalis, oils, and aliphatic hydrocarbons • Possesses self-extinguishing ability and high thermal stability • Good electrical conductivity • Exhibits high dimensional stability
[b]Polyvinylidene fluoride	$(C_2H_2F_2)n$	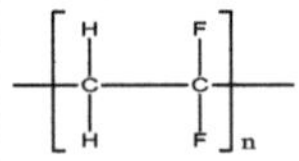	• Excellent resistance to impact, wear, and tear • Good electrical insulation properties • Exhibits piezoelectricity, enabling its use in sensors, transducers, and energy-harvesting devices • Highly resistant to UV radiation and weathering • Hydrophobic and resistant chemicals, acids, and solvents

(continued)

Table 2 (continued)

Polymer matrix	Chemical formula	Structure of monomer unit	Key properties of polymers
[b]Polyaniline	$(C_6H_7N)n$	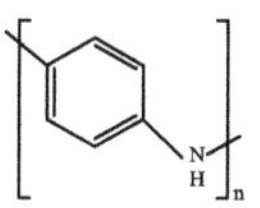	• Limited mechanical strength • High electrical conductivity • Good environmental stability • Good redox and electroactivity properties
[a]Epoxy	$(R_1COCR_2R_3)n$	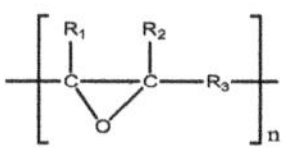	• Strong adhesive properties • High mechanical strength • Resistant to many chemicals, solvents, acids, and bases • High thermal stability • Good electrical insulation properties

[b]denotes thermoplastics; [a]denotes thermosets

5.1 *Mechanical Properties of the Polymers*

The mechanical behaviour of polymers is strongly dependent on their molecular weight, degree of crystallinity, crosslink density, and chain entanglement. The tensile strength and modulus increase with increasing molecular weight due to greater chain entanglement, which impedes slippage under stress [26]. Similarly, elasticity is characteristic of loosely cross-linked elastomers, where networked chains can deform and recover. Toughness is enhanced by energy dissipation mechanisms such as crazing or rubber toughening in multiphase systems [8].

Crystalline polymers, such as polyethylene terephthalate (PET), exhibit high stiffness due to ordered lamellar regions that resist deformation. In contrast, amorphous polymers such as polystyrene are more brittle and cannot reinforce crystalline domains [27].

5.1.1 Response to Applied Forces

Different kinds of mechanical stressors cause different reactions in polymers. A key characteristic that establishes how much force a polymer can bear before breaking is its tensile strength or the resistance of the material to stretching [28]. Polyethylene and polypropylene are examples of thermoplastics that frequently exhibit both exceptional ductility and high tensile strength [29]. Conversely, thermosetting polymers, such as epoxy resins, are more brittle but have a higher tensile strength [30].

The ability of a polymer to withstand pressures intended to compress or squeeze the material is indicated by its compressive strength [31]. Although filled or reinforced composites can perform noticeably better in compression, polymers typically exhibit lower compressive strength than metals do [32].

In load-bearing structural applications, the flexural strength of a material or its capacity to withstand deformation during bending is crucial [33]. It includes elements of compressive and tensile strength and offers information on how a polymer responds to bending stress [34]. Torsional strength, or the ability to withstand twisting, is also crucial in applications involving rotating forces, such as polymer composite drive shafts or gears [34].

5.1.2 Deformation Behaviour

When subjected to mechanical loads, polymers display distinct deformation behaviours. The ability of a polymer to regain its original shape in the absence of stress is known as elasticity [35]. Elongation at break, which indicates ductility, and Young's modulus, which quantifies stiffness, are used to quantify elongation. Rubber and other high-elasticity polymers can bend significantly before reverting to their original shape [36].

Nevertheless, viscoelasticity, which is a time-dependent mechanical behaviour that combines viscous and elastic responses, is also exhibited by many polymers. Owing to their dual nature, polymers are susceptible to changes in temperature, loading rate, and duration of applied stress [36]. Stress relaxation, in which the stress within a material diminishes under a constant strain, and creep, which is the slow deformation under a constant load over time, are two crucial viscoelastic phenomena [37]. These characteristics are particularly crucial for long-term applications where mechanical integrity must be preserved over time, such as gaskets, seals, and biomedical implants [38].

5.1.3 Resistance to Fracture

Another important mechanical factor is the way polymers fracture. One of the most important characteristics of impact resistance is toughness, which is the capacity of a material to absorb energy before breaking [39]. The temperature, strain rate, and existence of stress concentrators such as voids or notches are some of the variables that affect toughness. High-toughness polymers, such as polycarbonate, are frequently found in bulletproof materials and safety glasses [40].

Fatigue resistance, which characterises how polymers function under repeated loading and unloading cycles, is another crucial factor [41]. Cyclic stress has the potential to cause failure by causing cracks to form and spread over time. This is especially important for dynamic applications, including consumer electronics, prosthetic devices, and vehicle parts [42]. Although reinforcing can greatly improve performance, the natural capacity of polymers to disperse energy through their viscoelastic nature frequently results in greater fatigue resistance than metals do [43].

5.2 Surface Properties

Finally, the wear resistance and aesthetics depend on the surface characteristics of the polymers, particularly hardness. The term "hardness" describes the ability of a material to withstand scratches and surface indentations [44]. High-density polyethylene, for example, is prone to surface damage because of its comparatively low hardness, whereas polymers such as polyamide (nylon) or polyetheretherketone (PEEK) provide greater endurance and surface hardness [45]. Under abrasive conditions, the service life of polymer components can be extended by increasing their hardness with coatings or additives [46].

5.3 Thermal Properties of the Polymers

The glass transition temperature (Tg) and melting temperature (Tm) are central thermal transitions in polymers and are governed by chain flexibility, side-group size, and intermolecular interactions. Flexible backbones (e.g., poly(dimethylsiloxane)) yield a low Tg, whereas rigid or aromatic structures (e.g., polyimides) increase the Tg due to restricted segmental motion [47]. The melting temperature depends on the crystallinity and regularity of chain packing. For example, isotactic polypropylene crystallizes effectively and has a distinct Tm (~160 °C), whereas atactic variants remain amorphous and lack a defined melting point [8]. The thermal conductivity, which is typically low in polymers, can be increased through chain alignment or the incorporation of thermally conductive fillers such as boron nitride or graphene [48].

The glass transition temperature (Tg) is one of the most important thermal properties of polymers. The amorphous portions of a polymer change from a hard, glassy state to a soft, rubbery state at this temperature [49]. Brittleness results from the rigidity and restricted mobility of polymer chains below the Tg. The polymer becomes more elastic and flexible above the Tg because of enhanced molecular mobility [50]. For example, rubbery materials such as silicone have a significantly lower Tg and remain flexible at ambient temperature, but polystyrene has a Tg of over 100 °C and is hence hard [51].

The melting temperature (Tm), which is relevant to semicrystalline polymers, is another crucial thermal characteristic. The moment at which the crystalline sections of a polymer disintegrate and turn into a viscous liquid is known as the Tm [52]. Tm represents a real phase transition and is necessary for procedures such as injection moulding and extrusion, in contrast to Tg, which impacts amorphous domains. For example, the processing window for polyethylene is defined by its Tm, which is approximately 130 °C [53].

The temperature at which a polymer cools from an amorphous, disordered state to an ordered, crystalline form is known as the crystallisation temperature (Tc). The final degree of crystallinity, which affects the mechanical strength, optical clarity, and barrier qualities, is largely determined by the Tc [54].

In comparison with metals or ceramics, polymers often have low thermal conductivity, which means that they are not very good heat conductors [55]. This qualifies them for use in construction materials or electrical housing, for example, where thermal insulation is crucial. This also implies that heat dissipates gradually during processing, which may have an impact on the number of cycles and cooling rates [56].

The tendency of polymers to expand or shrink in response to temperature variations is known as thermal expansion [57]. In applications requiring tight tolerances or thermal cycling, polymers often have greater coefficients of thermal expansion than metals do. Mechanical failure or dimensional instability may result from excessive heat expansion [58].

Another crucial characteristic is heat capacity, which reflects how much energy is needed to increase the temperature of the polymer [59]. Higher heat capacity polymers require more energy to heat, which has an impact on thermal regulation and processing efficiency in electronics and packaging applications [60].

The ability of a polymer to withstand chemical deterioration at high temperatures is known as its thermal stability. In high-temperature applications such as cookware or automotive parts, this is crucial. The remarkable heat stability of polymers such as polytetrafluoroethylene (PTFE) and polyimides is well known [61].

Another major concern is flammability, or the capacity of a polymer to ignite and maintain combustion, particularly in sectors where safety is a top priority [62]. Although flame retardants and other chemicals can be applied to increase fire resistance, many polymers are naturally combustible [63]. Knowledge of the flammability and combustion byproducts of a polymer is crucial for fulfilling safety regulations in electronics, transportation, and construction.

5.4 Chemical Properties of the Polymers

5.4.1 Reactivity and Degradation

Depending on their structure and surroundings, polymers can undergo a variety of chemical reactions. Chemical degradation, which includes the disintegration of polymer chains as a result of processes such as oxidation and hydrolysis, is an important component [64]. Usually, hydrolysis takes place when water or moisture is present, particularly in polymers that have amide or ester bonds. Polyesters such as polylactic acid (PLA), for example, can break down in the presence of moisture, which qualifies them for biodegradable uses [64, 65]. Conversely, oxidation occurs when polymers react with oxygen, which is frequently sped up by heat or ultraviolet radiation [66]. Discolouration, loss of mechanical qualities, and chain scission are possible outcomes of this process. For example, unless stabilised with antioxidants, polyethylene, and polypropylene are susceptible to oxidative deterioration, particularly in outdoor settings [67]. Reactions that cross-link are yet another crucial chemical characteristic. Covalent bonds are created between polymer chains during cross-linking, creating a three-dimensional network [68]. This improves the mechanical, chemical, and thermal stability of the polymer. Extensive cross-linking gives thermosetting polymers, such as epoxy and vulcanised rubber, their strength and stiffness and renders them impervious to deformation and solvents [69].

Chemical reactivity can also be influenced by interactions between functional groups inside the polymer structure or with outside molecules. Compatibility with additives, fillers, or other polymers may be impacted by these interactions, which may also encourage or prevent deterioration [70].

Biodegradability describes the capacity of a polymer to break down under the influence of enzymes, microbes, or the environment [71]. Sustainable substitutes for traditional plastics are biodegradable polymers such as PLA and polyhydroxyalkanoates (PHAs) [72]. Chemical structure plays a major role in biodegradability, with natural-origin monomers and hydrolysable linkages being more vulnerable to microbial attack [73].

5.4.2 Interactions with Solvents

Different interactions between polymers and solvents are important in applications such as adhesives, coatings, drug delivery systems, and membrane technologies. The capacity of a polymer to dissolve in a particular solvent is referred to as solubility [74]. The chemical compatibility of the solvent and polymer, which is commonly described by the "like dissolves like" concept, determines this. In contrast to water, polystyrene dissolves in organic solvents such as toluene [75]. Many polymers, even those that are not completely soluble, exhibit swelling behaviour, in which the solvent permeates the matrix of the polymer and causes it to expand [76]. In applications such as hydrogels and sealants, swelling is essential because it alters the mechanical characteristics, permeability, and appearance of the polymer. The temperature, cross-link density, and polymer–solvent affinity all affect how much swelling occurs [77].

5.4.3 Barrier, Diffusion, and Transport Properties

The capacity of liquids or gases to permeate a polymer is known as permeability. This characteristic is crucial for barrier films, medical equipment, and packaging (such as food storage and breathable wound dressings) [78]. While some polymers, such as polyvinylidene chloride (PVDC), have good barrier qualities, others, such as polyethylene, are relatively permeable to gases. The barrier properties of polymers against gases, vapours, and liquids are influenced by temperature, free volume, polymer crystallinity, and the molecular size of the permeant [79]. Dense packing and crystallinity reduce diffusivity by limiting penetrant pathways. For example, polyvinylidene chloride (PVDC) has excellent barrier properties because of its high polarity and crystallinity [27]. Polymers with flexible chains and high free volume, such as poly(dimethylsiloxane), exhibit high permeability due to enhanced segmental mobility. Multilayer structures and nanocomposites further tune barrier performance by creating tortuous paths that increase the diffusion length.

5.5 Optical Properties

Optical behaviour is largely determined by morphology, electronic structure, and molecular packing. Transparency in polymers arises when the refractive index is homogeneous and scattering centers such as crystallites or phase domains are absent. Amorphous polymers (e.g., PMMA) are highly transparent, whereas semicrystalline polymers (e.g., HDPE) tend to be opaque due to light scattering at the crystal-morphous interfaces [80]. For optoelectronic applications, conjugated polymers (e.g., polythiophenes or polyfluorenes) display tunable bandgaps and fluorescence, which are controlled by backbone planarity and substituent effects. Subtle structural modifications can shift the emission colour, improve charge transport, or enable multifunctional behaviour in light-emitting diodes and sensors [81].

6 Understanding the Structure–Property Relationships in Polymeric Materials

The term structure–property relationship in polymer science refers to the correlation between the molecular, supramolecular, and macroscopic structural characteristics of a polymer and its physical, mechanical, thermal, optical, and barrier properties. Understanding this relationship is central to the rational design of polymers for both conventional and advanced applications. Unlike small-molecule materials, polymers exhibit emergent behaviour owing to their high molecular weight, chain entanglement, and morphological diversity. These factors endow them with unique properties such as viscoelasticity, processability, and the ability to form complex hierarchical structures. A fundamental understanding of structure–property relationships thus allows scientists and engineers to tailor polymer performance for specific functions, from high-strength composites used in aerospace devices to responsive hydrogels in biomedical devices [8, 27]. The structure of a polymer can be described on multiple levels:

- **Primary structure**: chemical composition and monomer sequence
- **Secondary structure**: chain conformation, tacticity, and stereochemistry

- **Tertiary structure**: supramolecular organization such as crystallinity and phase separation
- **Quaternary or bulk structure**: macroscopic arrangements, including network architecture, crosslinking density, and blend morphology

Each structural level contributes distinctly to the material's overall behavior. For example, increasing the molecular weight typically enhances the tensile strength and toughness due to greater chain entanglement [26]. Similarly, changes in tacticity—such as switching from atactic to isotactic polypropylene—can shift the polymer from amorphous to semi-crystalline, thereby altering its mechanical and thermal properties [47].

Moreover, subtle modifications in chemical structure—such as the introduction of polar functional groups or varying side chain lengths—can significantly affect the glass transition temperature (Tg), hydrophobicity, or barrier properties [80]. These structure–property relationships are not just empirical observations but also form the basis for predictive material design—a principle increasingly supported by computational modelling and high-throughput experimentation.

6.1 Hierarchical Structure of Polymeric Materials

The physical behaviour and performance of polymeric materials are dictated not by a single structural feature but by a hierarchy of structural levels, each contributing uniquely to the material's macroscopic properties. This hierarchical organization spans from the molecular architecture of individual polymer chains to the supramolecular assemblies and bulk-scale arrangements that emerge during processing and use.

6.2 Molecular-Level Structure

At the most fundamental level, the polymer structure begins with the chemical composition and configuration of the chains. This includes the identity and arrangement of the monomer units, molecular weight, polydispersity index (PDI), and chain architecture (linear, branched, star, dendritic, etc.). Additionally, tacticity—the stereochemical arrangement of side groups along the chain—profoundly influences crystallinity and thermal behaviour. For example, isotactic polypropylene forms semi-crystalline structures with a high melting point, whereas atactic polypropylene remains amorphous and soft [8, 47].

Chain rigidity, governed by the rotational freedom of bonds and the presence of aromatic or bulky side groups, affects the glass transition temperature (Tg) and mechanical stiffness.

The molecular weight and distribution also play crucial roles; higher molecular weights generally increase the tensile strength and impact resistance due to enhanced chain entanglement and intermolecular interactions [26].

6.3 *Supramolecular and Mesoscale Structures*

Above the molecular level, supramolecular organization governs many of the bulk properties of polymers. This includes crystalline versus amorphous domains, microphase separation in block copolymers, and self-assembly in supramolecular polymers. The degree of crystallinity influences the optical transparency, mechanical strength, and barrier properties. Semi-crystalline polymers such as polyethylene exhibit a two-phase morphology: crystalline lamellae embedded within an amorphous matrix, resulting in a unique combination of stiffness and ductility [27, 81].

In copolymers and polymer blends, phase separation and interfacial compatibility are central to determining morphology. Block copolymers, for instance, undergo microphase separation into ordered domains such as lamellae, cylinders, or spheres depending on block length ratios and interactions—structures that significantly affect mechanical, thermal, and transport properties [82, 83].

6.4 *Bulk Level and Network Structure*

At the macroscopic scale, the polymer architecture includes entanglements, crosslinks, and network connectivity, which define properties such as elasticity, toughness, and dimensional stability. Thermosets, such as epoxy resins, possess a high crosslink density that yields rigid, insoluble networks, whereas elastomers such as poly(dimethylsiloxane) (PDMS) maintain flexibility owing to the low density of soft crosslinks [80].

In polymer composites and nanocomposites, the distribution and orientation of filler particles (e.g., carbon nanotubes, clay, or graphene) within the polymer matrix introduce another layer of structural complexity. These systems exhibit property enhancements derived not only from the matrix polymer but also from the synergistic effects of hierarchical interactions across all scales [84].

6.5 *Interconnectivity of Structural Scales*

Critically, these structural levels do not operate independently; rather, they are interdependent and synergistic. Processing methods such as extrusion, solvent casting, and electrospinning influence molecular orientation and supramolecular ordering, which in turn dictate final material performance. A comprehensive understanding of this hierarchy enables precise control over properties such as permeability, mechanical strength, and thermal resistance, which are capabilities that are central to advanced polymer design [56].

7 Challenges and Future Directions in Structure–Property Engineering

The field of polymer design through structure–property relationships continues to evolve rapidly, with significant advances in materials synthesis, characterization techniques, and computational modelling. However, several challenges remain that hinder the realization of fully optimized polymeric materials for high-performance applications. This section outlines these challenges and discusses potential future directions that could address them.

7.1 Challenges in Predicting Structure–Property Relationships

While significant progress has been made in understanding how polymer structure affects properties, predictive modelling remains a complex and often imprecise task. Polymers exhibit a wide range of structural features, from simple linear chains to complex cross-linked networks, and these features can vary significantly at multiple length scales (molecular, nanoscopic, and macroscopic).

The heterogeneity in polymers, such as polydispersity in molecular weight or complex blending of different monomers, complicates efforts to establish direct correlations between structure and properties [85]. Furthermore, environmental factors such as temperature, humidity, and mechanical loading also play critical roles in determining material behaviour, but these variables are difficult to incorporate into predictive models.

7.2 Scaling from the Laboratory to Industrial Manufacturing

Although laboratory-scale syntheses of advanced polymers, such as block copolymers, nanocomposites, and high-performance polyimides, have yielded promising results, scalability remains a significant challenge. The techniques used in research often do not translate easily to industrial-scale processes due to issues with batch-to-batch variability, reproducibility, and cost-effectiveness [86].

In particular, achieving consistent nanofiller dispersion in large-scale processes such as extrusion or injection moulding continues to be problematic. The melting behavior of a polymer, which is sensitive to its molecular architecture, becomes difficult to control during scaling. Additionally, the lack of standardization in processing conditions and characterization methods limits the development of universally applicable models for scaling.

7.3 *Sustainability and Recycling of Polymers*

As the environmental impact of plastics becomes an increasing concern, one of the most pressing challenges is improving the sustainability of polymeric materials. The inherent non-biodegradability of many common polymers, such as polyethylene (PE) and polystyrene (PS), contributes significantly to pollution, particularly in the marine environment. Moreover, many polymers rely on non-renewable petrochemical feedstocks.

Recent research has focused on bio-based polymers and biodegradable plastics, such as polylactic acid (PLA) and polyhydroxyalkanoate (PHA). However, there are trade-offs between biodegradability, mechanical properties, and processing conditions that require careful optimization. Similarly, the recycling of polymers, especially thermosets and composite materials, remains a major challenge due to the difficulty of breaking down crosslinked structures without compromising material integrity [87].

7.4 *Advanced Computational and Machine Learning Approaches*

Computational materials science has advanced dramatically in recent years, particularly with the rise of machine learning (ML) and artificial intelligence (AI) techniques. These tools can rapidly analyse large datasets and uncover hidden patterns in the structure–property relationships of polymers. For instance, genetic algorithms have been applied to identify optimal polymer compositions for specific applications [88].

However, the accuracy and transferability of these models are still limited by the availability of high-quality data. Additionally, the complexity of polymer behavior means that purely data-driven approaches often lack the interpretability and physical insight provided by traditional modelling. Bridging the gap between empirical data and fundamental physics-based models will be crucial for enabling reliable predictions in material design [89].

7.5 *Future Directions: Smart Polymers and Functional Materials*

The development of smart polymers and functional materials will drive innovations across a range of industries, from biomedical applications to flexible electronics. Stimulus-responsive polymers, which can change their properties in response to external triggers such as light, heat, pH, or magnetic fields, are poised to revolutionize areas such as drug delivery, sensors, and actuators [90].

Advances in self-healing polymers, which can autonomously repair damage after mechanical failure, have shown promise in extending the lifespan and durability of polymeric materials. Additionally, polymers with integrated functionality, such as conducting polymers for wearable electronics or high-energy–density polymers for batteries, are likely to experience increased development in the coming years. The next frontier in polymer science will involve the integration of multiple functional properties into single polymer systems for diverse applications, from biomedical devices to autonomous repair and energy storage.

8 Conclusion

The study of structure–property relationships in polymers is not only fundamental to advancing materials science but also plays a pivotal role in driving technological innovation across various fields. The interdisciplinary nature of polymer science will continue to be a critical factor in addressing the material challenges of the future. As researchers deepen their understanding of the intricate connections between polymer structure and properties, the next generation of smart materials and sustainable solutions will emerge, offering tremendous benefits for society and the environment. The future of polymer science holds great promise, where well-designed polymers will not only enhance existing technologies but also open new avenues for advanced materials that meet the evolving needs of the twenty-first century.

References

1. Akharame, M.O., Fatoki, O.S., Opeolu, B.O., Olorunfemi, D.I., Oputu, O.U.: Polym.-Plast. Technol. Eng. **57**, 1801–1827 (2018)
2. Akharame, M.O., Oputu, O.U., Pereao, O., Fagbayigbo, B.O., Razanamahandry, L.C., Opeolu, B.O., Fatoki, O.S.: Nanostructured Materials for Treating Aquatic Pollution. Springer, pp. 275–306. (2019)
3. Saldívar-Guerra, E., Vivaldo-Lima, E.: Handbook of Polymer Synthesis, Characterization, and Processing, pp. 1–14. (2013)
4. Desidery, L., Lanotte, M.: Plastic Waste for Sustainable Asphalt Roads, pp. 3–28. Elsevier (2022)

5. Carraher Jr, C.E., Seymour, R.: Structure-Property Relationships in Polymers. Springer Science & Business Media (2012)
6. Tao, L., Byrnes, J., Varshney, V., Li, Y.: Iscience **25** (2022)
7. Vale, A.C., Leite, L., Pais, V., Bessa, J., Cunha, F., Fangueiro, R.: Polymers **16**, 3602 (2024)
8. Young, R.J., Lovell, P.A.: Introduction to Polymers. CRC Press (2011)
9. Tajeddin, B., Arabkhedri, M.: In: AlMaadeed, M.A.A., Ponnamma, D., Carignano, M.A. (eds.) Polymer Science and Innovative Applications, pp. 525–543. Elsevier (2020)
10. Zargar, V., Asghari, M., Dashti, A.: ChemBioEng Rev. **2**, 204–226 (2015)
11. Rajeswari, S., Prasanthi, T., Sudha, N., Swain, R.P., Panda, S., Goka, V.: World J. Pharm. Pharm. Sci. **6**, 472–494 (2017)
12. Irimia-Vladu, M., Sariciftci, N.S.: Polym. Int. **74**, 71–86 (2025)
13. Patel, R., Trivedi, R., Raj, M., Raj, L.: Chem. Pap. **78**, 665–697 (2024)
14. Nwuzor, I.C., Ogah, A.O., Ezika, A.C., Madu, I.O., Ibenta, M.E.: Hybrid Polymeric Systems for Biomedical Applications, pp. 53–89. Elsevier (2025)
15. Islam, S., Karim, F.E., Islam, M.R., Saeed, M.A., Alam, K.A., Khatun, M.M.: SPE Polym. **6**, e10157 (2025)
16. Scott, C.E., Macosko, C.W.: Polymer **36**, 461–470 (1995)
17. Xu, Y., Yao, X., Zhang, Z., Zhu, D., Na, H., Jin, Z., Fang, C.: Mater. Today Commun. 111500 (2025)
18. Nag, A., Arifianti, A.E., Khankhuean, A., Ajiro, H.: Europ. Polym. J. 113146 (2024)
19. Toh, H.W., Toong, D.W.Y., Ng, J.C.K., Ow, V., Lu, S., Tan, L.P., Wong, P.E.H., Venkatraman, S., Huang, Y., Ang, H.Y.: Eur. Polymer J. **146**, 110249 (2021)
20. Skosana, S.J., Khoathane, C., Malwela, T.: J. Thermoplast. Compos. Mater. **38**, 754–780 (2025)
21. Atta, M.R., Algethami, N., Farea, M.O., Alsulami, Q.A., Rajeh, A.: Int. J. Energy Res. **46**, 8020–8029 (2022)
22. Qu, K., Yuan, Z., Wang, Y., Song, Z., Gong, X., Zhao, Y., Mu, Q., Zhan, Q., Xu, W., Wang, L.: ChemPhysMater **1**, 294–309 (2022)
23. Callister, W.D., Rethwisch, D.G.: Materials Science and Engineering: an Introduction. John Wiley & Sons (2020)
24. Fried, J.R.: Polymer Science and Technology. Pearson Education (2014)
25. Venkatram, S., McCollum, J., Stingelin, N., Brettmann, B.: Polym. Int. **72**, 855–860 (2023)
26. Billmeyer, F.W.: Textbook of Polymer Science. John Wiley & Sons (1984)
27. Sperling, L.H.: Introduction to Physical Polymer Science. John Wiley & Sons (2015)
28. Machello, C., Bazli, M., Santos, J., Rajabipour, A., Arashpour, M., Hassanli, R.: Constr. Build. Mater. **438**, 137150 (2024)
29. Tsioptsias, C., Leontiadis, K., Tzimpilis, E., Tsivintzelis, I.: J. Plast. Film Sheet. **37**, 283–311 (2021)
30. Mi, X., Liang, N., Xu, H., Wu, J., Jiang, Y., Nie, B., Zhang, D.: Prog. Mater Sci. **130**, 100977 (2022)
31. Xiong, Z., Lin, L., Qiao, S., Li, L., Li, Y., He, S., Li, Z., Liu, F., Chen, Y.: J. Build. Eng. **47**, 103828 (2022)
32. Naseem, Z., Shamsaei, E., Sagoe-Crentsil, K., Duan, W.: Cem. Concr. Res. **158**, 106843 (2022)
33. Gani, A., Ibrahim, M., Ulmi, F., Farhan, A.: Res. Mater. **21**, 100534 (2024)
34. Gad, M.M., Fouda, S.M.: J. Prosthodont. **32**, 96–110 (2023)
35. Mayumi, K., Liu, C., Yasuda, Y., Ito, K.: Gels **7**, 91 (2021)
36. Yang, L., Yang, L., Lowe, R.L.: Mech. Mater. **157**, 103839 (2021)
37. Giuliani, P.M., Giannini, O., Panciroli, R.: Compos. Struct. **310**, 116755 (2023)
38. Schiffmann, K.I.: Int. J. Mater. Res. **97**, 1199–1211 (2022)
39. Sharafi, S., Santare, M., Gerdes, J., Advani, S.: Addit. Manuf. **38**, 101830 (2021)
40. Sharma, P., Mali, H.S., Dixit, A.: Iran. Polym. J. **30**, 193–233 (2021)
41. Guo, R., Li, C., Niu, Y., Xian, G.: J. Mater. Res. Technol. **21**, 4773–4789 (2022)
42. Zhang, Z., Chen, G., Xue, Y., Duan, Q., Liang, X., Lin, T., Wu, Z., Tan, Y., Zhao, Q., Zheng, W.: Adv. Func. Mater. **33**, 2305705 (2023)
43. Han, Z., Lu, Y., Qu, S.: Adv. Func. Mater. **34**, 2313498 (2024)

44. Al-Qahtani, A.S., Tulbah, H.I., Binhasan, M., Abbasi, M.S., Ahmed, N., Shabib, S., Farooq, I., Aldahian, N., Nisar, S.S., Tanveer, S.A.: Polymers **13**, 4077 (2021)
45. Yunisa, F., Effendi, A.C.S., Yasmin, M., Waskitho, A., Ashrin, M.N.: Surface roughness of polyetheretherketone and thermoplastic nylon after soaking in tea solution and mechanical cleaning. In: AIP Conference Proceedings, p. 020016. AIP Publishing LLC (2025)
46. Jain, M., Patil, S.: Mater. Today Proc. **63**, 40–45 (2022)
47. Odian, G.: Principles of Polymerization. John Wiley & Sons (2004)
48. Ren, J., Li, Q., Yan, L., Jia, L., Huang, X., Zhao, L., Ran, Q., Fu, M.: Mater. Des. **191**, 108663 (2020)
49. Tao, L., Varshney, V., Li, Y.: J. Chem. Inf. Model. **61**, 5395–5413 (2021)
50. Chen, G., Tao, L., Li, Y.: Polymers **13**, 1898 (2021)
51. Yan, J., Xu, J., Weng, L.-T., Wang, F., Wang, X., Yuan, H., Wang, T., Tsui, O.K.: Macromolecules **56**, 556–566 (2022)
52. Zou, L., Zhang, W.: Macromolecules **55**, 4899–4906 (2022)
53. Jia, Y., Mao, Z., Huang, W., Zhang, J.: Mater. Chem. Phys. **287**, 126325 (2022)
54. Kondo, M., Nishizono, T., Kawatsuki, N.: Mol. Cryst. Liq. Cryst. **733**, 9–15 (2022)
55. Zha, J. W., Wang, F., Wan, B.: Prog. Mater. Sci. 101362 (2024)
56. Li, R.-Y., Yang, X., Li, J., Shen, Y.-N., Zhang, L.-X., Lu, R., Wang, C., Zheng, X.-H., Chen, H.-S., Zhang, T.: Mater. Today Phys. **22**, 100594 (2022)
57. Colón Quintana, J.L., Slattery, L., Pinkham, J., Keaton, J., Lopez-Anido, R.A., Sharp, K.: Materials **15**, 2764 (2022)
58. Idumah, C.I.: Polym.-Plast. Technol. Mater. **62**, 1178–1203 (2023)
59. Fialko, N., Dinzhos, R., Sherenkovskii, J., Meranova, N., Aloshko, S., Izvorska, D., Korzhyk, V., Lazarenko, M., Mankus, I., Nedbaievska, L.: Eastern-Europ. J. Enterp. Technol. **6**, 114 (2021)
60. Djordjevic, K.L., Milicevic, D., Galovic, S., Suljovrujic, E., Jacimovski, S., Furundzic, D., Popovic, M.: Int. J. Thermophys. **43**, 68 (2022)
61. Kareem, A.A., Rasheed, H.K., Nasir, E.M.: Polym. Bull. **79**, 6617–6626 (2022)
62. Shamsuri, A.A., Yusoff, M.Z.M., Abdan, K., Jamil, S.N.A.M.: e-Polymers **23**, 20230060
63. Zaripov, I.I., Vikhareva, I.N., Buylova, E.A., Berestova, T.V., Mazitova, A.K.: Nanotekhnologii v Stroitel'stve **14**, 156–161 (2022)
64. Salahudeen, S., Thiel, T.A., Mejía, E.: RSC Sustain. **2**, 1904–1929 (2024)
65. Kojima, T., Takeda, J., Song, Y., Yamamoto, K., Ikeda, Y.: Int. J. Pharm. **647**, 123504 (2023)
66. Nukovic, A., Hamrangsekachaee, M., Rajkumar, M., Wong, G., Tressler, E.R., Hashmi, S.M., Hatfield, S.M., Bencherif, S.A.: Mater. Today Bio **32**, 101743 (2025)
67. Zhao, Y., Long, J., Zhuang, P., Ji, Y., He, C., Wang, H.: J. Mater. Chem. C **10**, 1010–1016 (2022)
68. Liu, Y., Yu, Z., Wang, B., Li, P., Zhu, J., Ma, S.: Green Chem. **24**, 5691–5708 (2022)
69. Wan, X., Demir, B., An, M., Walsh, T.R., Yang, N.: Int. J. Heat Mass Transf. **180**, 121821 (2021)
70. Saad, F., Mosaad, M., Othman, H., Hassabo, A.G.: Egypt. J. Chem. **67**, 99–108 (2024)
71. Mukherjee, C., Varghese, D., Krishna, J., Boominathan, T., Rakeshkumar, R., Dineshkumar, S., Rao, C.B., Sivaramakrishna, A.: Eur. Polymer J. **192**, 112068 (2023)
72. Samir, A., Ashour, F.H., Hakim, A.A., Bassyouni, M.: Npj Mater. Degrad. **6**, 68 (2022)
73. Baidurah, S.: Polymers **14**, 4928 (2022)
74. Marshall, J.E., Zhenova, A., Roberts, S., Petchey, T., Zhu, P., Dancer, C.E., McElroy, C.R., Kendrick, E., Goodship, V.: Polymers **13**, 1354 (2021)
75. Zhou, P., Sánchez-Rivera, K.L., Huber, G.W., Van Lehn, R.C.: Chemsuschem **14**, 4307–4316 (2021)
76. Zhan, Y., Fu, W., Xing, Y., Ma, X., Chen, C.: Mater. Sci. Eng., C **127**, 112208 (2021)
77. Metze, F.K., Sant, S., Meng, Z., Klok, H.-A., Kaur, K.: Langmuir: ACS J. Surf. Coll. **39**, 3546–3557 (2023)
78. Prasad, K., Nikzad, M., Sbarski, I.: Polym. Rev. **61**, 194–237 (2021)

79. Kanesugi, H., Ohyama, K., Fujiwara, H., Nishimura, S.: Int. J. Hydrogen Energy **48**, 723–739 (2023)
80. Chanda, M., Roy, S.K.: Plastics Technology Handbook. CRC Press (2006)
81. Zhao, Q., Qi, H.J., Xie, T.: Prog. Polym. Sci. **49**, 79–120 (2015)
82. Mai, Y., Eisenberg, A.: Chem. Soc. Rev. **41**, 5969–5985 (2012)
83. Bates, C.M., Bates, F.S.: Macromolecules **50**, 3–22 (2017)
84. Bayer, I.S.: Materials **10**, 748 (2017)
85. Zhou, D., Xu, M., Tan, R., Sun, Y., Ma, Z., Li, J., Dong, X.-H.: Polymer **225**, 123746 (2021)
86. Murphy, W., Black, J., Hastings, G.W.: Handbook of Biomaterial Properties. Springer (2016)
87. Utekar, S., Suriya, V., More, N., Rao, A.: Compos. B Eng. **207**, 108596 (2021)
88. Mitra, K.: Int. Mater. Rev. **53**, 275–297 (2008)
89. Wang, G., Wang, C., Zhang, X., Li, Z., Zhou, J., Sun, Z.: Iscience (2024)
90. Rosario, A.J., Ma, B.: ACS Appl. Polym. Mater. **6**, 14204–14228 (2024)

AI-Driven Optimization of Polymer Structure

Shakti Prasad Padhy, Jiaojiao Wang, Anshul Baral, and Raj Kiran

Abstract The design and optimization of polymer materials have traditionally relied on empirical and trial-and-error approaches, which are increasingly inadequate for navigating the vast chemical design space of macromolecules. This chapter presents a comprehensive overview of emerging artificial intelligence (AI)-driven strategies for polymer structure optimization, highlighting a paradigm shift toward data-driven and inverse design methodologies. Key aspects include advanced feature representation techniques for polymers, spanning molecular to mesoscale descriptors, and the integration of modern machine learning frameworks such as artificial neural networks, convolutional neural networks, graph neural networks, and transformer-based models for accurate property prediction. The chapter further explores generative modeling approaches—including variational autoencoders, generative adversarial networks, and diffusion models—for the inverse design of novel polymer structures with targeted functionalities. In addition, optimization algorithms such as Bayesian optimization, genetic algorithms, and particle swarm optimization are discussed for efficient exploration of high-dimensional design spaces and experimental planning. A detailed case study on AIoptimized conductive polymers demonstrates how closed-loop, self-driving laboratories enable autonomous discovery and process optimization, bridging the gap between computational design and real-world manufacturing. Finally, the chapter outlines key challenges, including data scarcity, model interpretability, and integration with experimental workflows, and discusses future directions toward human-in-the-loop systems and industrial-scale deployment. Overall, this work provides a unified framework for leveraging AI to accelerate polymer discovery, design, and commercialization.

S. P. Padhy (✉)
J. Mike Walker '66 Department of Mechanical Engineering, Texas A&M University, College Station, TX, USA
e-mail: padhy.shaktiprasad@gmail.com

J. Wang · A. Baral
School of Materials Science and Engineering, Nanyang Technological University, Singapore, Singapore

R. Kiran
School of Mechanical and Materials Engineering, Indian Institute of Technology Mandi, Mandi, India

B. Sharma et al. (eds.), *AI-Driven Polymer Design*, Engineering Materials,
https://doi.org/10.1007/978-981-95-8353-9_3

Keywords Polymer informatics · Inverse materials design · Machine learning for materials · Self-driving laboratories · Artificial intelligence in polymer science

1 Overview of Polymer Structure Optimization

The field of polymer science has traditionally relied on an empirical research paradigm, largely based on trial-and-error experimentation. This classical approach, while foundational, is inherently time-consuming, resource-intensive, and often characterized by low success rates [1]. The fundamental challenge facing modern material scientists is one of scale; the chemical space of potential carbon-based molecules is estimated to exceed 10^{60} distinct structures, rendering exhaustive exploration through conventional means not just impractical, but statistically impossible [2]. Navigating this immense landscape through traditional methods, which involve the painstaking synthesis and measuring of macroscopic properties one sample at a time, provides only a limited ability to probe the vast design space. Furthermore, the intrinsic complexity of polymer systems, characterized by compositional polydispersity, molecular weight distributions, and a strong coupling between processing conditions and final properties, has long anchored the field in conventional methodologies [3].

In recent years, a paradigm shift has occurred. The convergence of increasing computational power, advanced algorithms, and a growing trend toward open material databases has established a new paradigm for polymer research [4–6]. This approach is not merely a technological enhancement but a fundamental necessity in response to the geometric scale of the chemical design problem. AI-driven strategies offer a transformative framework for efficiently navigating this complex landscape, providing the ability to vastly augment the scale and precision of materials research. By shifting from an experience-driven model to one centered on data, AI is making the previously intractable "inverse design problem" tractable, where a desired target property is used to directly generate a candidate material structure [7, 8]. This pivot enables the rapid development of high-performance materials tailored for diverse applications in fields ranging from aerospace and electronics to energy storage and biomedical engineering.

2 Role of AI in Polymer Design and Development

AI, encompassing machine learning (ML) and deep learning (DL), is emerging as a powerful tool for addressing scientific and technical challenges across various disciplines, including polymer science. AI-driven methods are revolutionizing materials discovery and development by enabling data-driven predictions, uncovering hidden patterns, and automating labor-intensive tasks [9]. These tools are particularly effective in polymer science because they can efficiently identify latent relationships within high-dimensional, nonlinear, and multivariable processing–structure–property–performance (PSPP) spaces [10].

By learning complex, nonlinear mappings between inputs (e.g., chemical descriptors, processing conditions) and outputs (e.g., material properties), a well-trained AI model can uncover underlying physical or chemical principles. Despite its demonstrated potential, the rapid rise of AI has been disruptive, and many scientific fields were not fully prepared for its integration [11]. The adoption of AI in the polymer community has faced significant barriers, including technical complexity and a notable lack of interdisciplinary training [12]. A prevalent misconception is that AI is a "magic solution", which can lead to a misunderstanding of its unique capabilities and limitations. Additionally, the "black box" nature of many deep learning models can hinder trust and adoption in a field where understanding fundamental mechanisms is paramount [13]. The path to full integration, therefore, depends not only on continued technological innovation but also on fostering interdisciplinary dialogue and bridging the educational gaps that separate conventional polymer research from data-driven innovation.

3 Feature Representation of Polymer Structures

At the foundation of every AI model lies one critical step: *Feature Representation.* For small molecules, this task is relatively straightforward using notations such as Simplified Molecular-Input Line-Entry System (SMILES). Polymers, however, are characterized by repeating units of variable length, heterogeneous architectures, stereoregularity, branching, and crosslinking. Thus, a robust feature representation for polymers must achieve two objectives: simplification (reducing the infinite conformational space) and fidelity (preserving essential chemical information).

3.1 Hierarchical Representation of Polymers

One of the most powerful ways to understand polymer representation in AI models is to view it as a hierarchy of structural complexity, as shown in Fig. 1.

1. *Small Molecules and Monomers*: The foundation layer, encoded using Simplified Molecular Input Line Entry System (SMILES) or International Chemical Identifier (InChI) (e.g., Ethylene: C=C, Styrene: C=Cc1ccccc1, Vinyl chloride: C=CCl, etc.). To ensure generative robustness, the field has adopted PSELFIES (Pseudo-polymer Self-Referencing Embedded Strings), a graph grammar where every possible string corresponds to a valid molecular graph [14].
2. *Repeat Units and Stochasticity*: The repeat unit condenses the backbone structure. BigSMILES addresses the challenge of dispersity and random copolymers by introducing a "Stochastic Object" enclosed in curly brackets {} [15]. Instead of

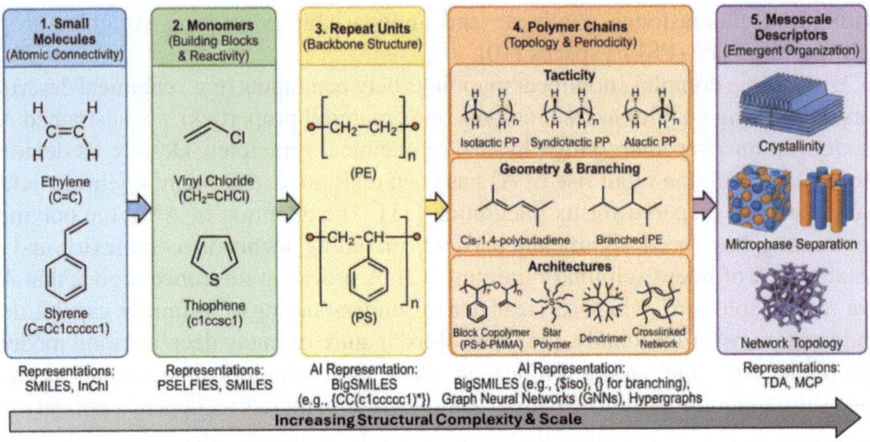

Fig. 1 Hierarchical feature representation of polymer structures for AI models

a fixed graph, it defines the repeating units and bonding rules (e.g., {[<]CC[>]} for polyethylene).

3. *Polymer Chains (Topology)*: Features like tacticity and cis/trans geometry govern crystallinity. AI representations encode these using stereoregularity tags (e.g., {$iso}, {$syn}) in BigSMILES or by propagating stereochemical information across GNNs. To solve the "infinite chain" problem, architectures like polyGNN create Cyclic Graphs that connect virtual nodes across boundaries [16].
4. *Mesoscale Descriptors*: At the highest level, emergent features like crystallinity and phase separation arise. Traditional string encodings fail here. Instead, Topological Data Analysis (TDA) extracts persistent features (loops, voids) from polymer configurations using persistence barcodes [17]. Multi-Cover Persistence (MCP) extends this by encoding higher-order structural info, outperforming standard fingerprints for heterogeneous morphologies [18].

4 Algorithms for Polymer Design

The computational design of polymers has evolved from simple empirical correlations to sophisticated data-driven frameworks capable of navigating the immense chemical space of macromolecules. Unlike small molecules, polymers exhibit hierarchical complexity which is defined not just by their atomic connectivity (primary structure) but also by their chain conformation, dispersity, phase morphology, and processing history. Addressing this complexity requires a diverse algorithmic toolkit. This section categorizes these algorithms into three distinct pillars: Deep Learning for property prediction, Generative Models for inverse molecular design, and Optimization Algorithms for experimental planning.

4.1 Deep Learning Approaches in Polymer Design

Deep Learning (DL) models, built on neural network architectures, serve as high-fidelity surrogates for physical experiments or quantum mechanical calculations. They learn the complex, non-linear mappings between molecular structure (or microstructure) and macroscopic properties. By learning representations directly from data, DL models can capture subtle, non-linear correlations that escape traditional empirical models.

4.2 Artificial Neural Networks (ANNs)

The Artificial Neural Network (ANN), specifically the Multi-Layer Perceptron (MLP), is the foundational architecture of deep learning. It consists of layers of neurons that mathematically transform input features (descriptors) into an output prediction. In polymer informatics, ANNs remains the workhorse of Quantitative Structure–Property Relationship (QSPR) modeling. Typically, ANNs take tabular data as input, where polymers are converted into numerical descriptors such as molecular weight, number of rotatable bonds, or counted fingerprints (e.g., the number of hydroxyl groups). The network minimizes a loss function, such as Mean Squared Error (MSE), to learn the mapping between these descriptors and a target property.

By processing input features through multiple layers of non-linear activation functions, ANNs can model complex response surfaces where variables interact in unpredictable ways. ANNs have become a standard for predicting the glass transition temperature (T_g). Recently, Fatriansyah et al. demonstrated that while tree-based models are robust, ANNs achieve superior accuracy ($R^2 = 0.79$) when trained on a large dataset (NIMS Polymer database PoLyInfo) because they better capture the non-linear interactions between functional groups and chain rigidity [19].

A compelling recent application of ANNs is found in the optimization of conductive polymer electrodes for supercapacitors. Alam and Husain benchmarked several machine learning algorithms to predict the specific capacitance of electrodes based on Polyaniline (PANI), Polypyrrole (PPy), and Polythiophene (PTh) [20]. Their study found that the MLP model achieved the highest predictive accuracy ($R^2 = 0.9622$) with the lowest Mean Absolute Error (MAE), significantly outperforming Decision Tree and Support Vector Machine (SVM) models. The primary weakness of ANNs is their reliance on hand-crafted descriptors; if a feature (such as long-range connectivity) is not explicitly calculated and fed to the model, the ANN cannot learn it.

4.3 Convolutional Neural Networks (CNNs)

While ANNs handle tabular data, Convolutional Neural Networks (CNNs) are specialized for processing grid-structured data, making them indispensable for analyzing polymer microstructure and morphology. CNNs employ learnable filters (kernels) that slide across an input image to generate feature maps, progressively detecting local patterns ranging from simple geometric features like edges and pores to complex hierarchical structures such as crystalline domains.

This architecture allows the model to automatically learn spatial hierarchies of features, a capability that Goh et al. leveraged to establish the Chemception framework [21]. Chemception demonstrated that deep CNNs could predict chemical properties using only 2D drawings of molecules converted into grid images, matching expert-developed QSPR models without requiring explicit chemical knowledge. Building on this concept, researchers have effectively adapted CNNs for polymer science by treating chemical and physical structural data as spatially structured "images". For example, Miccio and Schwartz demonstrated that polymer chemical structures could be represented as binary matrices derived from SMILES strings; these were processed by a CNN to accurately predict T_g without requiring manual feature engineering [22].

Addressing the more complex topology of crosslinked networks, Yan et al. utilized a dual-convolutional framework incorporating BigSMILES notation to discover new thermoset shape memory polymers with optimized recovery stress [23]. Moving beyond molecular composition to physical morphology, Ma et al. applied CNNs to optimize electrospinning process parameters (such as voltage and concentration), thereby controlling the diameter and uniformity of PVDF nanofibers for advanced sensing applications [24].

4.4 Graph Neural Networks (GNNs)

Graph Neural Networks (GNNs) have emerged as the architecture of choice for molecular modeling because they represent molecules in their natural form as graphs $G = (V, E)$, where atoms constitute the nodes V and chemical bonds form the edges E [25]. Unlike grid-based CNNs, GNNs employ a "message passing" algorithm where information about local atomic environments is iteratively aggregated and updated across the graph. This approach allows the model to learn molecular fingerprints directly from the structure without manual descriptor engineering.

To address the specific challenges of polymer science, such as modeling infinite chains and complex stoichiometries, researchers have developed specialized GNN architectures. Gurnani et al. introduced polyGNN [16], which converts repeat units into periodic graphs by adding bonds between boundary atoms to accurately model infinite chains. This architecture, combined with multitask learning, enabled the rapid screening of over 13,000 polymers across more than 30 properties. For more heterogeneous systems like polyesters, Queen et al. developed POLYMERGNN [26], which employs a unique pooling mechanism that processes acid and glycol monomers separately before concatenating their embeddings with resin properties.

Recent advancements have also focused on overcoming data scarcity. Volgin et al. tackled the "reality gap" by pre-training a GNN on an enormous synthetic dataset of over 6 million polyimides, significantly reducing error rates when fine-tuned on smaller experimental datasets [27]. Similarly, Gao et al. utilized self-supervised learning to learn molecular syntax by masking specific nodes or edges, reducing prediction errors by nearly 30% in low-data regimes [28].

4.5 Transformers and Large Language Models (LLMs)

Transformers and Large Language Models (LLMs) have revolutionized polymer informatics by treating polymer sequences as "chemical text", enabling the application of powerful Natural Language Processing (NLP) techniques to materials science. Central to these models is the self-attention mechanism [29], which allows the network to weigh the importance of different atoms in a sequence relative to one another, effectively capturing long-range dependencies.

Pioneering models like polyBERT [30] trained on 100 million polymer structures, leveraged this to create dense numerical fingerprints for rapid property prediction, outstripping traditional methods in speed by two orders of magnitude while preserving accuracy. Similarly, TransPolymer [31] demonstrated that transformer-based architectures with chemically aware tokenizers could outperform Recurrent Neural Networks (RNNs) in predicting properties like conductivity and crystallization tendency.

However, representing a complex macromolecule as a simple 1D string has limitations, driving a shift towards Multimodal Transformers that fuse distinct data modalities. Han et al. introduced an architecture combining 1D SMILES sequences with 2D molecular graph embeddings [32]. Furthermore, Vuong et al. recently expanded this multimodal frontier by fine-tuning Vision-Language Models (VLMs) with instruction-tuning pairs, demonstrating that combining visual data with text descriptions outperforms unimodal approaches [33].

5 Generative Models for Inverse Design

Generative models address the "inverse design" challenge by inverting the traditional predictive workflow, effectively mapping desired target properties back to optimized molecular structures. Instead of screening fixed libraries, these architectures navigate the continuous latent space of chemical possibilities to construct novel macromolecules with specific functionalities.

5.1 Variational Autoencoders (VAEs)

Variational Autoencoders (VAEs), originally introduced by Kingma and Welling [34], were among the first generative models applied to polymers. VAEs operate by mapping discrete molecular structures into a continuous, low-dimensional latent space. The architecture consists of an encoder that compresses the molecule into a probabilistic latent vector and a decoder that attempts to reconstruct it. Because this latent space is continuous (unlike the discrete nature of atoms and bonds), researchers can perform gradient-based optimization within it; moving along a specific vector corresponds to smoothly modifying the molecular structure to maximize a property.

Recent research has tailored VAEs to address specific challenges in polymer informatics. To generate chemically valid structures, Batra et al. implemented a Syntax-Directed VAE (SD-VAE) that constrains the decoder to follow grammatical rules, enabling the discovery of polymers with high T_g and bandgaps [35]. Similarly, Gurnani et al. introduced polyG2G, a novel graph-to-graph translation method that maps low-performing polymer graphs to high-performing ones, effectively designing dielectric polymers with high energy density [36]. Addressing topological complexity, Zheng et al. developed a graph-based VAE with dual encoders to design recyclable vitrimers [37].

5.2 Generative Adversarial Networks (GANs)

Generative Adversarial Networks (GANs), originally proposed by Goodfellow et al. [38], offer a competitive approach to generative design consisting of two neural networks: a generator (G) that creates synthetic structures and a discriminator (D) that evaluates their realism against real data. In this minimax game, G learns to produce increasingly realistic molecular graphs or SMILES strings to "fool" D.

However, applying standard GANs to polymers is challenging because chemical representations are discrete, preventing standard gradient backpropagation. To resolve this, researchers developed the Objective-Reinforced GAN (ORGAN) framework, which integrates Reinforcement Learning (RL) to treat the discriminator's assessment as a "reward" signal rather than a differentiable loss [39]. This allows the model to optimize not just for validity but for specific material properties. This was showcased by utilizing ORGAN framework to bias the generation distribution toward specific physicochemical metrics and successfully optimize candidates for organic photovoltaics [40]. Additionally, Liu et al. used GANs for data augmentation on Solution Styrene-Butadiene Rubber (SSBR), generating high-fidelity synthetic samples that improved T_g prediction accuracy ($R^2 > 0.95$) despite limited experimental data [41].

5.3 Diffusion Models

Diffusion models have rapidly emerged as the state-of-the-art in generative AI, offering a robust alternative to VAEs and GANs for high-fidelity molecular design. The process involves two stages: a *forward diffusion process*, where structure is gradually destroyed by adding Gaussian noise until the data becomes pure randomness, and a learnable *reverse diffusion process*, where a neural network is trained to iteratively "denoise" the random noise back into a coherent sample [42].

In polymer science, this iterative refinement is particularly advantageous for generating complex 3D atomic structures and molecular graphs. Unlike GANs, which often suffer from mode collapse (generating repetitive samples), diffusion models provide stable training and high diversity. Jain et al. introduced polyGen, the first latent diffusion transformer designed explicitly for atomic-level polymer generation [43]. By combining graph encodings with positional biased attention, polyGen generates realistic 3D crystal structures from minimal repeat unit inputs. Similarly, Yang et al. applied diffusion models to the de novo design of solid polymer electrolytes, identifying 17 novel candidates with ionic conductivities significantly higher than existing benchmarks [44].

6 Optimization Algorithms for Experimental Planning

While deep learning models serve as high-fidelity surrogates to map the polymer chemical space, optimization algorithms act as the navigation engine, guiding the search for superior materials and optimal processing conditions.

6.1 Bayesian Optimization (BO)

Bayesian Optimization (BO) has established itself as the standard for "active learning" in materials science, particularly for problems where data is sparse and experiments are expensive. BO is a sequential design strategy that does not merely predict properties but intelligently selects the next experiment to perform. It operates by constructing a probabilistic surrogate model, typically a Gaussian Process (GP), that maps the chemical design space to a target property, providing both a predicted value (mean) and an uncertainty estimate (variance). An acquisition function (e.g., Expected Improvement) then evaluates this surface to select candidates that maximize utility, balancing the trade-off between exploitation (improving upon known high-performers) and exploration (probing uncertain regions).

Gao et al. utilized a machine-learning-assisted BO framework to revolutionize the design of polyamide nanofiltration membranes [45]. By screening an "infinite" space of monomer combinations, they identified novel formulations that broke the traditional trade-off between permeability and salt rejection. Similarly, Ruža et al. demonstrated a fully autonomous, closed-loop laboratory for polymer electrolytes using a "warm-start" batch BO strategy to identify poly(ethylene oxide) (PEO)-based formulations with ionic conductivities exceeding 10^{-4} S/cm [46].

6.2 Genetic Algorithm (GA)

Genetic Algorithms (GA) are a class of evolutionary algorithms inspired by the principles of natural selection. In the context of polymer informatics, GAs treat molecular structures or processing parameters as "chromosomes" composed of discrete "genes". The algorithm evolves a population of candidate solutions through an iterative process of selection—crossover and mutation. A fitness function, often a machine learning surrogate, evaluates each generation.

Because GAs do not require gradient information, they are uniquely suited for the discrete, combinatorial, and often non-differentiable landscape of polymer chemistry. Kim et al. demonstrated this by designing polymers for extreme conditions; they utilized a GA to assemble chemical fragments into polymer chains, successfully identifying candidates with a T_g exceeding 500 K and a bandgap greater than 6 eV [47]. To improve efficiency, Patra et al. proposed a Neural-Network-Biased GA, where a neural network predicts which mutations are most likely to yield improvements [48].

6.3 Particle Swarm Optimization (PSO)

Particle Swarm Optimization (PSO) is a stochastic population-based metaheuristic inspired by the social behavior of bird flocking and fish schooling [49]. In PSO, a group of candidate solutions (particles) moves through the search space to find the global optimum. Each particle adjusts its trajectory based on its own best-known position and the best position found by the entire swarm.

PSO is computationally efficient and effective for continuous, non-linear optimization problems in polymer engineering. Moayyedian et al. applied Multi-Objective PSO to the injection molding of automotive dashboards, optimizing cooling time and melting temperature to minimize defects [50]. Beyond processing, PSO is increasingly used for molecular self-assembly. Tsai and Fredrickson pioneered the PSO-SCFT method [51], coupling PSO with Self-Consistent Field Theory to discover globally stable morphologies for complex block copolymers.

7 Case Study: AI-Optimized Conductive Polymers

7.1 *Problem Definition*

Conductive polymers, such as PEDOT:PSS (Poly(3,4-ethylenedioxythiophene) polystyrene sulfonate), are foundational to flexible biosensors and wearable electronics due to their ability to be solution-processed into conformable thin films. However, optimizing these materials involves navigating a "trilemma" of intrinsically coupled and often conflicting performance requirements:

1. *Electrical conductivity* (σ): Achieving high σ typically requires maximizing carrier density and improving microstructural pathways through high crystallinity or chain planarity [52, 53].
2. *Seebeck Coefficient* (S): In organic thermoelectrics, strategies that increase carrier concentration often degrade S, limiting the overall Power Factor ($PF = S^2\sigma$) [54].
3. *Mechanical flexibility*: Processing routes that enhance conductivity, such as aggressive doping, rigid additives, or densification, frequently embrittle the material, making it unsuitable for flexible applications [55].

This optimization challenge is high-dimensional and non-linear. Even with a fixed chemistry like PEDOT:PSS, the design space includes formulation ratios, additive concentrations, and annealing parameters. For instance, the autonomous Polybot system faced a 7-dimensional search space exceeding 933,000 potential combinations, rendering traditional trial-and-error impractical [56]. The objective is not to find a single "best" polymer but to discover a Pareto set of solutions that balance these trade-offs.

7.2 *Phase 1: Predictive Modeling*

Predictive modeling for conductive polymers must account for the fact that electronic properties (band gap, ionization potential) are governed not just by monomer chemistry but by processing history and mesoscale organization. While classical Machine Learning (ML) using handcrafted descriptors (e.g., Random Forest, Gaussian Process Regression) can identify high σ regimes, it is often limited by poor transferability to new chemistries.

To address this, the field is shifting toward Deep Learning (DL) and representation learning:

1. *Transformer Models*: Architectures like TransPolymer [31] use chemically informed tokenization to deliver strong accuracy across diverse benchmarks, while prompt-based models like PolyNC [57] unify natural language inputs with chemical strings for flexible prediction.
2. *GNNs*: Multitask GNNs learn features directly from repeat units, sharing representations across related properties to improve efficiency [16].

3. *Self-Supervised Learning*: Pretraining on large unlabeled datasets allows models to learn robust electronic descriptors even in data-scarce regimes. For example, pretraining has been shown to reduce prediction errors for electron affinity and ionization potential by over 19% [28].

7.3 Phase 2: Generative Design

Generative design is emerging as a practical engine to translate predictive models into actionable recipes, operating through two parallel routes:

1. *Molecular-Space Optimization*: DL-based generators propose new polymer architectures. PolyTAO uses a Transformer-based directed pretraining strategy on a massive dataset of ~1 million structure–property pairs to achieve over 99% chemical validity [58]. Recently, LLM-based and diffusion-based models have also identified novel polymer electrolytes that outperform existing benchmarks [44].
2. *Process-Space Optimization*: This route uses Bayesian Optimization (BO) to navigate high-dimensional manufacturing variables. The Polybot platform exemplifies this by optimizing solution processing (formulation and coating) to produce transparent films with conductivities exceeding 4500 S/cm [56].

Crucially, modern generative workflows integrate feasibility layers. Tools like Generative BigSMILES encode synthesis-relevant statistics [59], while the Open Macromolecular Genome (OMG) filters candidates to ensuring they remain within synthetically accessible chemical spaces [60].

7.4 Phase 3: Closing the Loop (Autonomous Synthesis)

The final and most critical phase of the optimization workflow involves embedding planning, execution, and learning within Self-Driving Laboratories (SDLs). In the context of conductive polymers, theoretical inverse design is insufficient because electronic properties (e.g., charge transport) are path-dependent, i.e., dictated not just by molecular chemistry but by the specific history of processing, defect formation, and mesoscale organization. Consequently, the optimization target must shift from a theoretical property value to a reproducible manufacturing recipe. This transition is justified by two state-of-the-art platforms that demonstrate how AI-driven autonomy resolves the "reality gap" in polymer processing (Fig. 2).

1. *Modular Autonomy for Functional Films (PANDA)*: The Polymer Analysis and Discovery Array (PANDA) validates the SDL approach for electrodeposited films, where performance depends heavily on real-time deposition parameters [61].

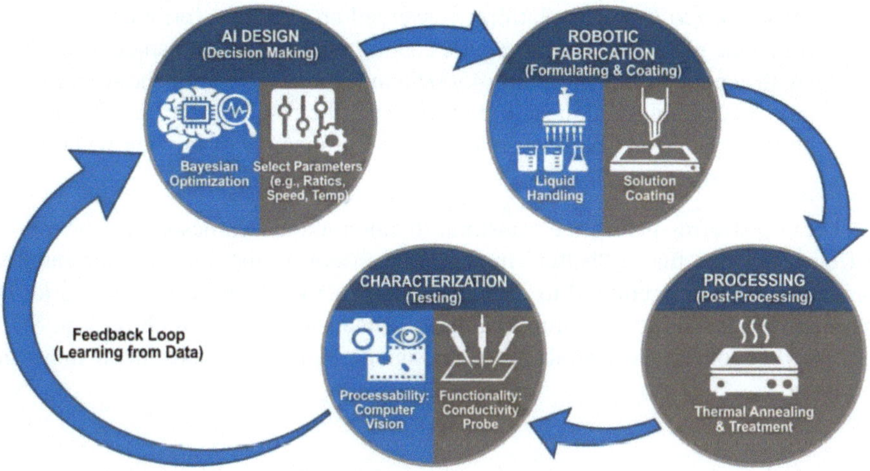

Fig. 2 The closed-loop workflow of a Self-Driving Laboratory, Polybot [56]

- *Workflow*: PANDA integrates a CNC-based liquid handling gantry with in-situ electrochemical and optical characterization. It executes a closed loop where a BO algorithm iteratively updates deposition voltage, time, and monomer concentration based on the measured electrochromic contrast of the resulting film.
- *Justification*: By automating the synthesize-characterize-learn cycle, PANDA rapidly identified deposition protocols that maximized electrochromic switching, achieving a model-predicted optimum that matched experimental observations within 20 iterations. This confirms that AI agents can effectively navigate the coupled chemical-electrical design space of functional polymers without human intervention.

2. *Process–Structure–Property Optimization (Polybot)*: While PANDA addresses electrodeposition, the Polybot platform provides the direct solution to the Conductivity-Flexibility-Processability trade-off central to this case study (Fig. 2). It targets solution-processed PEDOT:PSS, navigating a complex 7-dimensional search space (encompassing formulation additives, coating speeds, and post-processing thermal budgets) exceeding 933,000 potential combinations [56].

   - *The "Gatekeeper" Strategy*: A critical innovation in Polybot was the use of Importance-Guided BO. Standard algorithms might aggressively pursue high conductivity parameters that lead to film dewetting (defects). Polybot incorporated a computer vision module to quantify film coverage first, treating "processability" as a probabilistic gatekeeper. This ensured the AI learned to avoid recipes that were theoretically conductive but physically unmanufacturable.

- *From Lab to Fab*: The platform converged on a specific formulation (5 vol% ethylene glycol, high-temperature coating) that yielded transparent thin films with conductivities exceeding 4500 S/cm. Crucially, because the AI optimized for processability alongside performance, the resulting recipe was successfully transferred to a large-scale roll-to-roll manufacturing line without loss of performance.

These platforms justify the transition to autonomous synthesis: they prove that AI does not just find a "better" molecule; it discovers the specific, non-intuitive processing window required to make that molecule perform in the real world. By treating reproducibility and film quality as primary optimization objectives, SDLs like Polybot convert the abstract inverse design problem into an executable industrial protocol.

8 Summary and Conclusions

8.1 Key Takeaways and Industrial Implications

The integration of Artificial Intelligence into polymer science marks a definitive transition from the classical Edisonian paradigm, characterized by intuition and trial-and-error, to a regime of data-driven precision. This chapter has explored the computational infrastructure required to navigate the immense chemical space of polymers, highlighting that "inverse design" is no longer a theoretical aspiration but a practical reality.

The transformation rests on three algorithmic pillars:

1. ***Deep Learning Surrogates***: Predictive models, ranging from GNNs to Transformers like PolyBERT, now allow researchers to screen millions of candidates for properties like glass transition temperature (T_g) and bandgap with near-DFT accuracy, bypassing slow physical experimentation.
2. ***Generative Design***: The "inverse" problem is being solved by generative models such as PolyTAO and diffusion probabilistic models. These tools do not merely predict properties of known molecules but autonomously imagine novel chemical structures that meet specific performance targets, often achieving chemical validity rates exceeding 99%.
3. ***Closed-Loop Optimization***: As demonstrated by the conductive polymer case study, algorithms like BO are most powerful when coupled with robotic automation. These SDLs are not merely automated tools but orchestrated cyber-physical systems that manage the flow of data, matter, and decision-making logic simultaneously.

Industrial Implications: The implications for the polymer industry are profound. First, the efficiency of R&D is dramatically increased; autonomous platforms like Polybot have demonstrated the ability to converge on optimal formulations in fewer than 50 iterations, a fraction of the effort required by traditional methods. Second, the emergence of commercial "Cloud Labs" like Emerald Cloud Lab (ECL) allows companies to access this infrastructure without massive capital expenditure, shifting the business model from selling instruments to selling *Lab-as-a-Service* (*LaaS*). This democratization enables a "virtualization" of materials science, where digital design is instantly translatable to physical execution.

9 Challenges, Future Trends, and the Path of Commercialization

Despite the successes of SDLs in parameter optimization, the path to widespread commercial adoption faces significant hurdles regarding data, trust, and human integration.

9.1 The Challenge of Data Scarcity and Quality

A primary bottleneck in polymer AI is the "small data" problem. Unlike computer vision or natural language processing, where labeled data is abundant, polymer science suffers from sparse, heterogeneous, and often noisy experimental datasets.

1. ***Scarcity***: High-fidelity property data (e.g., experimental bandgaps or conductivity) is expensive to generate. To overcome this, strategies like transfer learning and pre-training are essential. For instance, PolyTAO leverages large-scale pre-training on fundamental properties to learn chemical syntax before fine-tuning on smaller downstream tasks, significantly improving performance in data-scarce regimes [58].
2. ***Quality and Standardization***: Data quality is as critical as quantity. Inconsistent reporting standards and "silent" parameters (e.g., unrecorded ambient humidity during coating) can lead to reproducibility crises. Future SDLs must enforce FAIR principles (Findable, Accessible, Interoperable, and Reusable) at the point of generation, ensuring that every data point captures the full context of its creation [62].

9.2 The "Black Box" and Trust

In industrial settings, understanding the *mechanism* of failure is often as critical as predicting success. The "black box" nature of deep learning agents hinders trust, as researchers may hesitate to rely on a system that cannot explain its decisions. Future systems must move beyond simple prediction to Hypothesis Learning, where the AI not only finds the optimum but proposes a physical mechanism or rule that explains why that optimum exists [63, 64].

9.3 Future Trend: From Automation to Human-in-the-Loop (HITL)

The most significant future trend is the shift from fully autonomous "lights-out" laboratories to Human-in-the-Loop (HITL) systems. As Hysmith et al. argue, removing the human entirely is often counterproductive because AI lacks the high-level reasoning, inductive bias, and ethical judgment of a scientist [65].

- **Interactive AI**: Future SDLs will function as collaborative partners rather than independent agents. The AI will handle high-dimensional optimization and routine execution, while the human provides strategic guidance, constraints, and "sanity checks" at critical decision points.
- **Gamification**: To facilitate this interaction, interfaces are evolving toward gamification. By wrapping complex optimization tasks in intuitive, game-like user interfaces (UIs), SDLs can lower the cognitive load for researchers. This approach can even enable crowdsourced decision-making, where multiple experts "play" the optimization game to guide the AI out of local minima.

9.4 The Path of Commercialization: From Lab-to-Fab and Democratization

The commercial viability of AI-driven polymers relies on transforming SDLs from academic prototypes into industrial engines. This transition is occurring through two parallel trajectories: the integration of manufacturing logic and the democratization of hardware.

The "Lab-to-Fab" Paradigm Shift: A major bottleneck in commercialization is the traditional 10–20 year timeline required to translate a laboratory discovery into a market-ready product. As Low et al. argue, current workflows are sequential, i.e., optimizing performance first and manufacturability later [66]. The path forward lies in "Manufacturing-Aware Discovery", where SDLs incorporate downstream constraints (e.g., toxicity, cost, supply chain availability, and scalable processing methods like Roll-to-Roll coating) into the initial discovery phase. By collapsing the distinction between materials screening and manufacturing optimization, next-generation SDLs will deliver recipes that are pre-validated for the factory floor, significantly de-risking commercial scale-up.

Democratization via Frugal Automation: While "Cloud Labs" like Emerald Cloud Lab offer centralized power, the high cost of commercial automation (often exceeding

$100,000) creates a technological divide, limiting access to well-funded institutions [64]. To counter this, a wave of "Frugal Automation" is emerging, driven by state-of-the-art instruments and open-source hardware [67]. This democratization is critical for commercialization because it allows startups and smaller labs to deploy custom SDLs rapidly, fostering a decentralized ecosystem of innovation alongside centralized cloud facilities.

Ultimately, the commercial future of polymer science will be defined by this hybrid ecosystem: centralized Cloud Labs handling massive, standardized throughput, while distributed, low-cost SDLs drive niche innovation. Both will operate on a Lab-to-Fab logic, ensuring that the AI-designed materials of the future are not just theoretically superior, but practically manufacturable.

References

1. Ge, W., De Silva, R., Fan, Y., Sisson, S.A., Stenzel, M.H.: Machine learning in polymer research. Adv. Mater. **37**(11), 2413695 (2025)
2. Dobson, C.M.: Chemical space and biology. Nature **432**(7019), 824–828 (2004)
3. Yue, T., He, J., Li, Y.: Machine-learning-assisted molecular design of innovative polymers. Acc. Mater. Res. **6**(8), 1033–1045 (2025)
4. Li, Y., Yue, T.: Machine Learning for Polymer Informatics (ACS in Focus). American Chemical Society, pp. 1 (2024)
5. Patra, T.K.: Data-driven methods for accelerating polymer design. ACS Polym. Au **2**(1), 8–26 (2022)
6. Jain, A., Ong, S.P., Hautier, G., Chen, W., Richards, W.D., Dacek, S., Cholia, S., Gunter, D., Skinner, D., Ceder, G., Persson, K.A.: Commentary: the materials project: a materials genome approach to accelerating materials innovation. APL Mater. **1**(1) (2013)
7. Wang, J., Wang, Y., Chen, Y.: Inverse design of materials by machine learning. Materials **15**(5), 1811 (2022)
8. Sattari, K., Xie, Y., Lin, J.: Data-driven algorithms for inverse design of polymers. Soft Matter **17**(33), 7607–7622 (2021). https://doi.org/10.1039/D1SM00725D
9. Madika, B., Saha, A., Kang, C., Buyantogtokh, B., Agar, J., Wolverton, C.M., Voorhees, P., Littlewood, P., Kalinin, S., Hong, S.: Artificial intelligence for materials discovery, development, and optimization. ACS Nano **19**(30), 27116–27158 (2025)
10. Xie, C., Qiu, H., Liu, L., You, Y., Li, H., Li, Y., Sun, Z., Lin, J., An, L.: Machine learning approaches in polymer science: progress and fundamental for a new paradigm. SmartMat **6**(1), e1320 (2025)
11. Khlaif, Z.N., Mousa, A., Hattab, M.K., Itmazi, J., Hassan, A.A., Sanmugam, M., Ayyoub, A.: The potential and concerns of using AI in scientific research: ChatGPT performance evaluation. JMIR Med Educ **9**, e47049 (2023)
12. Boyce, B., Dingreville, R., Desai, S., Walker, E., Shilt, T., Bassett, K.L., Wixom, R.R., Stebner, A.P., Arroyave, R., Hattrick-Simpers, J., Warren, J.A.: Machine learning for materials science: barriers to broader adoption. Matter **6**(5), 1320–1323 (2023)
13. Holm, E.A.: In defense of the black box. Science **364**(6435), 26–27 (2019)
14. Savit, A., Sahu, H., Shukla, S., Xiong, W., Ramprasad, R.: polyBART: a chemical linguist for polymer property prediction and generative design (2025). arXiv preprint arXiv:2506.04233
15. Lin, T.-S., Coley, C.W., Mochigase, H., Beech, H.K., Wang, W., Wang, Z., Woods, E., Craig, S.L., Johnson, J.A., Kalow, J.A., Jensen, K.F., Olsen, B.D.: BigSMILES: a structurally-based line notation for describing macromolecules. ACS Cent. Sci. **5**(9), 1523–1531 (2019)

16. Gurnani, R., Kuenneth, C., Toland, A., Ramprasad, R.: Polymer informatics at scale with multitask graph neural networks. Chem. Mater. **35**(4), 1560–1567 (2023)
17. Hiraoka, Y., Nakamura, T., Hirata, A., Escolar, E.G., Matsue, K., Nishiura, Y.: Hierarchical structures of amorphous solids characterized by persistent homology. Proc. Natl. Acad. Sci. **113**(26), 7035–7040 (2016)
18. Zhang, Y., Shen, C., Xia, K.: Multi-cover persistence (MCP)-based machine learning for polymer property prediction. Brief Bioinform. **25**(6), bbae465 (2024)
19. Fatriansyah, J.F., Linuwih, B.D., Andreano, Y., Sari, I.S., Federico, A., Anis, M., Surip, S.N., Jaafar, M.: Prediction of glass transition temperature of polymers using simple machine learning. Polymers **16**(17), 2464. https://doi.org/10.3390/polym16172464.
20. Alam, M., Husain, S.: Hyperparameter tuned machine learning predictions of specific capacitance of conducting polymers and their composites for high performance advanced supercapacitors. Appl. Phys. A **131**(1), 67 (2024)
21. Goh, G.B., Siegel, C., Vishnu, A., Hodas, N.O., Baker, N.: Chemception: a deep neural network with minimal chemistry knowledge matches the performance of expert-developed QSAR/QSPR models (2017). arXiv preprint arXiv:1706.06689
22. Miccio, L.A., Schwartz, G.A.: From chemical structure to quantitative polymer properties prediction through convolutional neural networks. Polymer **193**, 122341 (2020)
23. Yan, C., Feng, X., Wick, C., Peters, A., Li, G.: Machine learning assisted discovery of new thermoset shape memory polymers based on a small training dataset. Polymer **214**, 123351 (2021)
24. Ma, M., Jin, C., Yao, S., Li, N., Zhou, H., Dai, Z.: CNN-optimized electrospun TPE/PVDF nanofiber membranes for enhanced temperature and pressure sensing. Polymers **16**(17), 2423. https://doi.org/10.3390/polym16172423
25. Scarselli, F., Gori, M., Tsoi, A.C., Hagenbuchner, M., Monfardini, G.: The graph neural network model. IEEE Trans. Neural Netw. **20**(1), 61–80 (2009)
26. Queen, O., McCarver, G.A., Thatigotla, S., Abolins, B.P., Brown, C.L., Maroulas, V., Vogiatzis, K.D.: Polymer graph neural networks for multitask property learning. npj Comput. Mater. **9**(1), 90 (2023)
27. Volgin, I.V., Batyr, P.A., Matseevich, A.V., Dobrovskiy, A.Y., Andreeva, M.V., Nazarychev, V.M., Larin, S.V., Goikhman, M.Y., Vizilter, Y.V., Askadskii, A.A., Lyulin, S.V.: Machine learning with enormous "Synthetic" data sets: predicting glass transition temperature of polyimides using graph convolutional neural networks. ACS Omega **7**(48), 43678–43691 (2022)
28. Gao, Q., Dukker, T., Schweidtmann, A.M., Weber, J.M.: Self-supervised graph neural networks for polymer property prediction. Mol. Syst. Des. Eng. **9**(11), 1130–1143 (2024)
29. Vaswani, A., Shazeer, N., Parmar, N., Uszkoreit, J., Jones, L., Gomez, A.N., Kaiser, Ł., Polosukhin, I.: Attention is all you need. In: Presented at the Proceedings of the 31st International Conference on Neural Information Processing Systems, Long Beach, California, USA (2017)
30. Kuenneth, C., Ramprasad, R.: PolyBERT: a chemical language model to enable fully machine-driven ultrafast polymer informatics. Nat. Commun. **14**(1), 4099 (2023)
31. Xu, C., Wang, Y., Barati Farimani, A.: TransPolymer: a transformer-based language model for polymer property predictions. npj Comput. Mater. **9**(1), 64 (2023)
32. Han, S., Kang, Y., Park, H., Yi, J., Park, G., Kim, J.: Multimodal transformer for property prediction in polymers. ACS Appl. Mater. Interfaces **16**(13), 16853–16860 (2024)
33. Vuong, A., Van, M.-H., Verma, P., Zhao, C., Wu, X.: Fine-tuning vision-language models for multimodal polymer property prediction (2025). arXiv preprint arXiv:2511.05577
34. Kingma, D.P., Welling, M.: Auto-encoding variational Bayes (2013). arXiv preprint arXiv:1312.6114
35. Batra, R., Dai, H., Huan, T.D., Chen, L., Kim, C., Gutekunst, W.R., Song, L., Ramprasad, R.: Polymers for extreme conditions designed using syntax-directed variational autoencoders. Chem. Mater. **32**(24), 10489–10500 (2020)
36. Gurnani, R., Kamal, D., Tran, H., Sahu, H., Scharm, K., Ashraf, U., Ramprasad, R.: PolyG2G: a novel machine learning algorithm applied to the generative design of polymer dielectrics. Chem. Mater. **33**(17), 7008–7016 (2021)

37. Zheng, Y., Thakolkaran, P., Biswal, A.K., Smith, J.A., Lu, Z., Zheng, S., Nguyen, B.H., Kumar, S., Vashisth, A.: AI-guided inverse design and discovery of recyclable vitrimeric polymers. Advanced Science **12**(6), 2411385 (2025)
38. Goodfellow, I., Pouget-Abadie, J., Mirza, M., Xu, B., Warde-Farley, D., Ozair, S., Courville, A., Bengio, Y.: Generative adversarial networks. Commun. ACM **63**(11), 139–144 (2020)
39. Guimaraes, G.L., Sanchez-Lengeling, B., Outeiral, C., Farias, P.L.C., Aspuru-Guzik, A.: Objective-reinforced generative adversarial networks (organ) for sequence generation models (2017). arXiv preprint arXiv:1705.10843
40. Sanchez-Lengeling, B., Outeiral, C., Guimaraes, G.L., Aspuru-Guzik, A.: Optimizing distributions over molecular space. An objective-reinforced generative adversarial network for inverse-design chemistry (ORGANIC). (2017)
41. Liu, Z., Huo, Y., Chen, Q., Zhan, S., Li, Q., Zhao, Q., Cui, L., Liu, J.: Predicting the glass transition temperature of polymer based on generative adversarial networks and automated machine learning. Mater. Genome Eng. Adv. **2**(4), e78 (2024)
42. Ho, J., Jain, A., Abbeel, P.: Denoising diffusion probabilistic models. Adv. Neural Inf. Process. Syst. **33**, 6840–6851 (2020)
43. Jain, A., Srivastava, A., Ramprasad, R.: A learning framework for atomic-level polymer structure generation. Chem. Mater. **37**(18), 7337–7346 (2025)
44. Yang, Z., Ye, W., Lei, X., Schweigert, D., Kwon, H.-K., Khajeh, A.: De novo design of polymer electrolytes using GPT-based and diffusion-based generative models. npj Comput. Mater. **10**(1), 296 (2024)
45. Gao, H., Zhong, S., Zhang, W., Igou, T., Berger, E., Reid, E., Zhao, Y., Lambeth, D., Gan, L., Afolabi, M.A., Tong, Z., Lan, G., Chen, Y.: Revolutionizing membrane design using machine learning-Bayesian optimization. Environ. Sci. Technol. **56**(4), 2572–2581 (2022)
46. Ruža, J., Stolberg, M.A., Cawthern, S., Johnson, J.A., Shao-Horn, Y., Gómez-Bombarelli, R.: Autonomous discovery of polymer electrolyte formulations with warm-start batch Bayesian optimization. Chem. Mater. **37**(19), 7771–7777 (2025)
47. Kim, C., Batra, R., Chen, L., Tran, H., Ramprasad, R.: Polymer design using genetic algorithm and machine learning. Comput. Mater. Sci. **186**, 110067 (2021)
48. Patra, T.K., Meenakshisundaram, V., Hung, J.-H., Simmons, D.S.: Neural-network-biased genetic algorithms for materials design: evolutionary algorithms that learn. ACS Comb. Sci. **19**(2), 96–107 (2017)
49. Kennedy, J., Eberhart, R.: Particle swarm optimization. In: Proceedings of ICNN'95—International Conference on Neural Networks, 27 Nov.–1 Dec. 1995, vol. 4, pp. 1942–1948 (1995). https://doi.org/10.1109/ICNN.1995.488968
50. Moayyedian, M., Qazani, M.R.C., Amirkhizi, P.J., Asadi, H., Hedayati-Dezfooli, M.: Multiple objectives optimization of injection-moulding process for dashboard using soft computing and particle swarm optimization. Sci. Rep. **14**(1), 23767 (2024)
51. Tsai, C.L., Fredrickson, G.H.: Using particle swarm optimization and self-consistent field theory to discover globally stable morphologies of block copolymers. Macromolecules **55**(12), 5249–5262 (2022)
52. Shi, X.-L., Li, N.-H., Li, M., Chen, Z.-G.: Toward efficient thermoelectric materials and devices: advances, challenges, and opportunities. Chem. Rev. **125**(16), 7525–7724 (2025)
53. Shi, X.-L., Zou, J., Chen, Z.-G.: Advanced thermoelectric design: from materials and structures to devices. Chem. Rev. **120**(15), 7399–7515 (2020)
54. Bubnova, O., Khan, Z.U., Malti, A., Braun, S., Fahlman, M., Berggren, M., Crispin, X.: Optimization of the thermoelectric figure of merit in the conducting polymer poly(3,4-ethylenedioxythiophene). Nat. Mater. **10**(6), 429–433 (2011)
55. Lipomi, D.J., Tee, B.C., Vosgueritchian, M., Bao, Z.: Stretchable organic solar cells. Adv. Mater. (Weinheim) **23** (2011)
56. Wang, C., Kim, Y.-J., Vriza, A., Batra, R., Baskaran, A., Shan, N., Li, N., Darancet, P., Ward, L., Liu, Y.: Autonomous platform for solution processing of electronic polymers. Nat. Commun. **16**(1), 1498 (2025)

57. Qiu, H., Liu, L., Qiu, X., Dai, X., Ji, X., Sun, Z.-Y.: PolyNC: a natural and chemical language model for the prediction of unified polymer properties. Chem. Sci. **15**(2), 534–544 (2024)
58. Qiu, H., Sun, Z.-Y.: On-demand reverse design of polymers with PolyTAO. npj Comput. Mater. **10**(1), 273 (2024)
59. Schneider, L., Walsh, D., Olsen, B., de Pablo, J.: Generative BigSMILES: an extension for polymer informatics, computer simulations and ML/AI. Digit. Discov. **3**(1), 51–61 (2024)
60. Kim, S., Schroeder, C.M., Jackson, N.E.: Open macromolecular genome: generative design of synthetically accessible polymers. ACS Polym. Au **3**(4), 318–330 (2023)
61. Quinn, H., Robben, G.A., Zheng, Z., Gardner, A.L., Werner, J.G., Brown, K.A.: PANDA: a self-driving lab for studying electrodeposited polymer films. Mater. Horiz. **11**(21), 5331–5340 (2024)
62. Lo, S., Baird, S.G., Schrier, J., Blaiszik, B., Carson, N., Foster, I., Aguilar-Granda, A., Kalinin, S.V., Maruyama, B., Politi, M., Tran, H., Sparks, T.D., Aspuru-Guzik, A.: Review of low-cost self-driving laboratories in chemistry and materials science: the "frugal twin" concept. Digit. Discov. **3**(5), 842–868 (2024). https://doi.org/10.1039/D3DD00223C
63. Ziatdinov, M.A., Liu, Y., Morozovska, A.N., Eliseev, E.A., Zhang, X., Takeuchi, I., Kalinin, S.V.: Hypothesis learning in automated experiment: application to combinatorial materials libraries. Adv. Mater. **34**(20), 2201345 (2022)
64. Tobias, A.V., Wahab, A.: Autonomous 'self-driving' laboratories: a review of technology and policy implications. R. Soc. Open Sci. **12**(7), 250646 (2025)
65. Hysmith, H., Foadian, E., Padhy, S.P., Kalinin, S.V., Moore, R.G., Ovchinnikova, O.S., Ahmadi, M.: The future of self-driving laboratories: from human in the loop interactive AI to gamification. Digit. Discov. **3**(4), 621–636 (2024). https://doi.org/10.1039/D4DD00040D
66. Low, A.K.Y., Cheng, J.J.W., Hippalgaonkar, K., Ng, L.W.T.: Self-driving laboratories: translating materials science from laboratory to factory. ACS Omega **10**(28), 29902–29908 (2025)
67. Doloi, S., Das, M., Li, Y., Cho, Z.H., Xiao, X., Hanna, J.V., Osvaldo, M., Ng Wei Tat, L.: Democratizing self-driving labs: advances in low-cost 3D printing for laboratory automation. Digit. Discov. **4**(7), 1685–1721 (2025). https://doi.org/10.1039/D4DD00411F

Multiscale Modeling and AI Integration

Tanvi Yadav, Reetu Sharma, and Amit Kumar Sharma

Abstract From atomic scale to the bulk level, properties of polymers have been investigated for decades using classical modeling approaches such as quantum mechanics, molecular dynamics, coarse-grained models, finite element analysis, etc., however, these are constrained by scale integration and computational entities. Polymers are a large family of materials whose structure consists of several smaller molecules (called building blocks) and are used in many industries ranging from food packaging to aerospace, therefore, predicting their long-term performance is a grave challenge. Recent technological impacts on artificial intelligence and machine learning render them a novel method to analyze polymers differently. These innovative new approaches do not supplant current models but rather supplement them with data-driven and predictive tools which make design and discovery faster and reliable. This chapter is centered on addressing how the phase of artificial intelligence is applied to interpolating polymers at various scale levels through the simplest demonstration of how artificial intelligence could be integrated at the various scales. The case studies are also explained to enlighten how the AI-driven approaches lead to improve the prediction of important parameters such as glass transition temperature, blend miscibility, mechanical strength and degradation behavior. All in all, polymer modeling has reached an innovation era as it is becoming more predictive and reliable.

Keywords Multiscale modeling · Computational materials design · Materials informatics

T. Yadav · R. Sharma (✉)
Department of Chemistry, Netaji Subhas University of Technology, Dwarka, New Delhi, India
e-mail: ritus1066@gmail.com

A. K. Sharma
Department of Chemistry, Atma Ram Sanatan Dharma College, University of Delhi, Dhaula Kuan, New Delhi, India

B. Sharma et al. (eds.), *AI-Driven Polymer Design*, Engineering Materials,
https://doi.org/10.1007/978-981-95-8353-9_4

1 Introduction

Multiscale modeling is a substantiation harness employed in mathematical calculations to regulate the action of semi complex materials by interconnectedness of numerous amounts of characterization at different spatial and temporal resolutions. It is bridging the gap between the world of interatomic and the world of the bulk material properties. This is even better with polymers since the materials are already in the form of layers. Repeating units of atoms (monomers) form a polymer chain that is attached in a manner of chains and networks. However, these chains are mesogenic and form mesoscopic crystalline lamellae in the shape of amorphous or phase separated structures. Eventually, these structural features determine the shape of the microscopic property profile such as mechanical, optical and thermal properties in bulk materials [1]. Ab initio quantum mechanical molecular dynamics or molecular dynamic simulations give details about an interaction between specific atoms/bonds. These approaches offer detailed study of interatomic interactions, chain fold and chemical reactions, all limited to short time (femto seconds to nanosecond) and length (nanometers) scales. Couse-grained atom dynamics, or coarse-grained molecular dynamics mesoscopic simulations remove the finer particles and adds larger physics to the simulation over micro or milliseconds [2]. A continuum approach is regarded to be on the most basic level, to consider the utilization of polymers in their continuous phase by techniques such as finite element method (FEM). Such models may be employed to model macroscopic behavior like mechanical behavior of stress–strain, machinery crack progression, or long-term visco-selective creep [3]. The advantage of using the multiscale modeling is that it enables the connection of these different scales. High scale model parameters are estimated using lower scale information. Simulations in quantum mechanics which yield force fields can be applied in the study of the atomistic dynamics of molecules and statistical data on the atomistic particulars can be applied to enhance the coarse-grained models. Similarly, constitutive models at the continuum level are acquired with mesoscopic knowledge [4]. Combining these scales, scientist could predict influences on macroscopic characteristics of nanoscale chemistry and morphology. Such predictions would be either resource-demanding, or extremely incorrect if not for multiscale modeling.

1.1 Artificial Intelligence in Computational Modeling

Artificial intelligence is revolutionizing computational materials science, providing fresh approaches to accelerate predictions and detect hidden correlations. Traditionally, polymer modeling has relied on direct physical formulations and simulations grounded in first-principles or statistical mechanics. While these approaches are powerful, they often demand excessive computational resources for large systems or long timescales. AI offers an auxiliary method by studying directly from data, avoiding the need to simulate all atoms or time step explicitly [5]. AI spans numerous algorithms, beginning from foundational rule-based expert systems, which relied on manually encoded knowledge, progressing to high-level paradigms such as machine learning and deep learning. In Machine Learning, models are trained on datasets to correlate input and output relationship such as predicting glass transition temperatures based on molecular descriptors. Deep learning, with its layered neural network architectures, further broadened its capability to capture highly nonlinear and multifaceted relationships, such as predicting force fields directly from quantum mechanical energies [6].

For polymers, AI has become vital for several reasons:

- Instead of spanning millions of simulation steps at the atomistic level, AI models trained on smaller datasets is capable of forecasting behavior to much longer timescales [7].
- Certain phenomena, such as degradation, creep, or microphase separation, are challenging to model using physics-based models alone. AI models can recognize patterns in data from experiments or simulation that represent these behaviors [8].
- By mining databases of polymer properties, AI algorithms can predict unexplored structures for specific applications, such as high-strength composites, biodegradable packaging, or conductive films [9].

1.2 Scope and Objective of the Chapter

- We will explore how atomistic (molecular dynamics, quantum chemistry), mesoscopic (coarse-graining, DPD, Monte Carlo), and macroscopic (continuum mechanics, FEM) approaches come together to form a modeling hierarchy.
- The role of AI will be highlighted in force field development, coarse-graining, and property prediction, along with case studies that will show its capacity to stimulate polymer simulations.
- We will investigate how AI can provide broader span of predictions into timescales inaccessible by traditional methods, enabling forecasts of polymer aging, degradation, creep, and relaxation.
- Real-world applications, including blend miscibility, polyurethane foam degradation, and UV-induced degradation pathways will demonstrate the practical life impact of the above-mentioned methods.

2 Multiscale Modeling of Polymers

2.1 Atomistic Modeling: Molecular Dynamics and Quantum Mechanics

At the foundation of multiscale modeling lies atomistic modeling, which represents polymers at the resolution of atoms and chemical bonds. The two most widely used methods here are quantum mechanical simulations and molecular dynamics. Quantum mechanical approaches, particularly density functional theory and ab initio methods offering highly valid portrayal of electronic structures, bond formation, and chemical reactivity. They are indispensable for parameterizing force fields, calculating binding energies, or understanding photochemical degradation of polymers under UV irradiation. However, their computational cost restricts applications to small molecular fragments or oligomers, typically under a few hundred atoms [10, 11]. Molecular dynamics, in contrast, integrates Newton's equations of motion for thousands to millions of atoms over femtosecond timescales. Molecular Dynamic simulations allow investigation of polymer conformations, chain dynamics, crystallization, and local stress distributions [12]. Classical force fields such as COMPASS, OPLS-AA, AMBER, and CHARMM are widely used for polymers. Recent developments include reactive force fields (ReaxFF), which enable bond breaking and formation, thereby capturing degradation or curing reactions [13].

Atomistic modeling has provided key insights into:

- Glass transition temperature (T_g) prediction.
- Chain packing and entanglement density.
- Local segmental motions underlying viscoelasticity.
- Interfacial behavior in polymer nanocomposites.

2.2 Mesoscale Modeling: Coarse-Grained MD, DPD, and Monte Carlo

The mesoscale regime (10–1000 nm, μs–ms) connects the atomistic and continuum worlds. At this level, polymers are no longer represented by every atom, but instead by coarse-grained (CG) beads, each representing several atoms or monomers.

Coarse-Grained Molecular Dynamics (CGMD)

CGMD reduces degrees of freedom, allowing exploration of larger systems such as block copolymers, micelles, or phase-separated blends. The MARTINI force field is among the most successful CG frameworks, mapping 3–4 heavy atoms into one bead [14]. CG models are typically parameterized against atomistic simulations or experimental data.

Dissipative Particle Dynamics (DPD)

DPD is a mesoscopic technique particularly useful for soft matter. It introduces soft repulsive forces and random dissipative interactions between beads, allowing integration over longer time steps. DPD excels in capturing hydrodynamic interactions, microphase separation in block copolymers, and self-assembly processes [15].

Monte Carlo (MC) Methods

Stochastic sampling techniques such as configurational-bias Monte Carlo (CBMC) are often used to equilibrate dense polymer melts, where MD struggles due to slow dynamics. MC has proven powerful in studying chain conformations, adsorption on surfaces, and equilibrium morphologies.

2.3 Macroscale Modeling: Continuum Mechanics and FEM

At the macroscale (μm–m, seconds–years), polymers are treated as continuous media governed by constitutive equations. This level addresses engineering-scale phenomena such as stress–strain behavior, fracture, creep, and thermal transport. Continuum mechanics describes polymers using parameters such as elastic modulus, viscosity, and Poisson's ratio. Constitutive models like Maxwell, Kelvin-Voigt, or Generalized Viscoelastic models, are commonly used for viscoelastic polymers. The finite element method discretizes geometries into elements and solves partial differential equations describing deformation, diffusion, or heat transfer. FEM has been applied to:

- Predict crack propagation in polymer composites [16].
- Model viscoelastic creep under constant load.
- Simulate swelling in hydrogels under external stimuli.

2.4 Linking Across Scales

The essence of multiscale modeling lies in transferring information across scales:

Bottom-up approaches: Information flows from small to large scales. For example, force constants derived from QM feed into MD simulations, which in turn provide parameters (diffusion coefficients, entanglement length) for mesoscale models. These then parameterize constitutive equations at the continuum level.

Top-down approaches: Macroscopic experimental data are used to refine lower-scale models. For instance, stress–strain curves can be used to calibrate MD force fields or CG potentials.

Two primary strategies for linking scales are:

Sequential coupling: Each scale is simulated independently, and outputs are passed as inputs to the next scale.

Concurrent coupling: Multiple scales are simulated simultaneously, with dynamic feedback between them. Examples include QM/MM methods, where quantum chemistry is embedded in a molecular mechanics environment, or adaptive resolution MD, which switches between atomistic and coarse-grained resolution in different regions.

2.5 Case Studies in Multiscale Polymer Modeling

Polymer Blend Miscibility

Multiscale approaches combining atomistic MD and mesoscale DPD have predicted phase separation in blends of polymer, including polystyrene/poly(methyl methacrylate) (PS/PMMA), informing processing strategies for miscible blends [17].

Fracture Mechanics

Atomistic simulations reveal bond rupture mechanisms, which feed into FEM models predicting crack growth in polymer adhesives [18].

Viscoelasticity

Time–temperature superposition principles, combined with atomistic simulations of segmental relaxation, have informed constitutive models predicting creep and stress relaxation in epoxy resins [19].

Polymer Nanocomposites

Multiscale modeling has been used to understand reinforcement in polymer–clay nanocomposites. Atomistic MD captures polymer–filler interactions, mesoscale CGMD describes filler dispersion, and FEM predicts bulk modulus and toughness [20].

Hydrogels and Responsive Polymers

Coupled MD-FEM frameworks have modeled swelling, drug release, and mechanical response of stimuli-responsive hydrogels, guiding biomedical applications [21] (Table 1).

3 AI Integration in Multiscale Modeling

The last decade has observed a profound shift in computational materials science, largely due to the speedy growth of artificial intelligence. While traditional multiscale modeling frameworks remain irreplaceable, AI provides a supplementary framework

Table 1 Comparison of scales

Scale	Method	System size	Timescale	Output/properties
Atomistic	MD, DFT	100–1000 s atoms	fs–ns	Bond lengths, angles, interaction energies
Mesoscopic	CG, DPD, MONTE CARLO	10^3–10^6 beads	ns–μs	Phase separation, chain organization
Macroscopic	FEM, continuum models	mm–m	s–years	Stress–strain, creep, fracture behavior

that can boost the simulations, reduce computational costs, and reveal patterns unnoticed by physics-based model alone. In polymers, where phenomena occur at different scales of magnitude in space and time, this relation between artificial intelligence and multiscale modeling is especially compelling.

3.1 Machine Learning Algorithms in Polymer Simulation

Machine learning encompasses a group of algorithms that can learn complex relationships from data. Unlike physics-based models, that are based on direct mathematical formulations derived from fundamental principles, ML models approximate unknown functions directly from examples. This feature provides significant value in polymer science, where experimental or simulated datasets can be large and nonlinear, and where direct analytical relationships are hard to obtain.

Supervised Learning

In supervised ML, models are fitted using annotated data that pairs input descriptors (e.g., polymer chain length, side-group chemistry, molecular weight distribution) with output properties (e.g., glass transition temperature, tensile modulus). Once trained, the model enables property prediction for novel polymer systems. A prominent example is the prediction of the glass transition temperature (T_g). Traditionally, prediction of T_g is done using empirically derived relationships or computationally expensive molecular dynamics simulations. ML models, such as support vector machines (SVMs) or gradient boosting regressors, have proven capable for accurate T_g estimation using molecular descriptors derived from cheminformatics tools [22, 23]. This progresses the screening of expansive polymer datasets for specific applications.

Unsupervised Learning

Unsupervised ML does not require labeled outputs. Instead, it identifies patterns, clusters, or latent structures within datasets. For polymers, unsupervised methods prove as effective tools for analyzing macroscopic simulation or experimental datasets, such as microphase-separated block copolymer morphologies. For instance, clustering algorithms like k-means or Gaussian mixture models (GMMs) are used for classification of microstructures generated during polymer self-assembly [24]. Similarly, principal component analysis (PCA) or autoencoders can distill high-dimensional simulation outputs into a manageable latent space, revealing variations in polymer chains or phase diagrams.

Reinforcement Learning

Reinforcement learning is a latest methodology where an "agent" learns optimal strategies by interacting with an environment and uses reward-driven feedback. In polymer science, reinforcement learning has been used to refine reaction routes, polymer processing conditions, or additive formulations [25]. For example, the "agent" can sequentially suggest polymerization conditions to obtain high yield or tune annealing conditions to fulfill the intended phase-separated morphology.

3.2 *AI-Assisted Force Field Development*

One of the main difficulties in atomistic polymer simulations is the development of accurate force fields—mathematical functions that describe atomic interactions. Conventional force fields like AMBER, OPLS, or COMPASS rely on thorough calibration based on quantum mechanical calculations and experimental fitting, which is both time-consuming and possess limited scope. Artificial Intelligence has changed this field with machine learning potentials (MLPs), that are trained to give quantum mechanical correctness only at fraction of the computational cost. Neural network (NN) potentials, such as ANI [26] and DeePMD [27], learn directly from datasets of quantum energies and forces. Once trained, these models can predict interatomic interactions with near-DFT accuracy while permitting simulations over many orders of magnitude larger and longer.

For polymers, this is particularly important because:

1. Polymers often include heteroatoms, functional groups, and diverse bonding patterns. ML force fields can perform brilliantly across this range.
2. ML potentials can capture bond breaking and formation pathways that classical force fields often fail to represent.
3. Once given training, MLPs enable microsecond-scale molecular dynamics for bulk polymer chains.

3.3 *Data-Driven Coarse-Graining*

While atomistic simulations provide detailed insights, they are limited to nanometer scales and nanosecond times. To study mesoscale structures such as domain formation, crystallization, or blend miscibility, coarse graining (CG) is required. CG

Table 2 AI in multiscale polymer modeling

Method	Scale	Input data	Output	Application
NN potential	Atomistic	QM energies AND forces	Force fields	Faster MD with DFT-level accuracy
Regression CG	Mesoscopic	Atomistic configs	CG potentials	Phase separation prediction
Supervised ML	All	Polymer descriptors	Property predictions	T_g, modulus, conductivity
RL	Processing	Processing parameters	Optimized strategies	Annealing for morphology control
Unsupervised	Mesoscopic	Simulation snapshots	Microstructure maps	Block copolymer phase classification

reduces computational cost by grouping atoms into larger "beads" while preserving essential physics. Traditionally, CG potentials are derived by iterative Boltzmann inversion (IBI) or force matching, where atomistic simulation outputs are used to calibrate CG interactions. These methods, however, struggle with transferability across different conditions and require labor-intensive fine-tuning. AI has emerged as a valuable approach for automating coarse graining. Neural networks or regression models can learn the mapping from atomistic trajectories to effective CG potentials [28]. For example:

- Graph neural networks (GNNs) can represent polymer topologies and predict CG interactions directly.
- Autoencoders compress atomistic coordinates into latent representations that serve as CG beads.
- Kernel ridge regression (KRR) has been used to reproduce thermodynamic observables of polymer melts.

A notable success story is the reproduction of thermodynamic properties (e.g., pressure, radial distribution functions) of polymer melts using ML-derived CG potentials that outperform classical methods in transferability [29] (Table 2).

4 AI in Simulating Long-Time Behavior of Polymers

One of the central challenges in polymer science is the prediction of long-term behavior. Polymers in real-world applications, whether in packaging, automotive parts, medical devices, or structural composites, undergo changes that span years or decades. These include physical aging, chemical degradation, creep, and relaxation. Conventional molecular dynamics (MD) and continuum simulations cannot access such extended timescales, as atomistic simulations typically operate within nanoseconds to microseconds, while engineering lifetimes span months to decades. Artificial

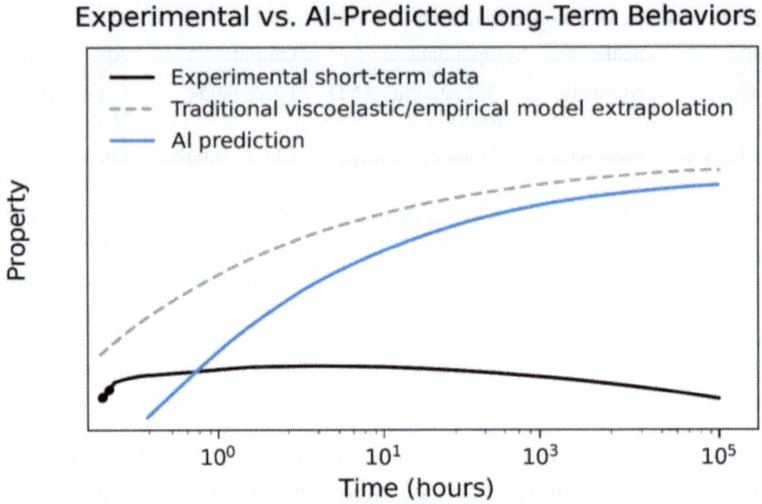

Fig. 1 A graph comparing experimental versus AI-predicted long-term behaviors. The x-axis is time (log scale, up to 10^5 h), and the y-axis is property (e.g., modulus, strain, degradation index)

intelligence offers a powerful solution: it can interpolate and extrapolate from short-term experimental or simulation data to predict long-term polymer performance. By recognizing patterns across datasets, AI models can forecast behaviors well beyond the reach of traditional methods (Fig. 1).

4.1 Aging of Polymers

Aging in polymers corresponds to the gradual change in physical and mechanical properties over time due to chain rearrangements, free volume reduction, and subtle molecular reorganizations. For instance, amorphous polymers below their glass transition temperature undergo physical aging, characterized by densification and a gradual increase in stiffness and brittleness [30]. Crosslinking or oxidative processes can further complicate aging dynamics. Classical models, such as the Kohlrausch–Williams–Watts (KWW) stretched exponential function, describe relaxation phenomena but require extensive experimental data and lack predictive power outside measured ranges. Atomistic MD simulations cannot extend to the years-long timescales relevant to physical aging. AI, particularly neural networks and Gaussian process regression, has been applied to predict the evolution of mechanical properties over extended times. For example:

- Neural networks trained on accelerated aging datasets (obtained via high-temperature or high-stress experiments) can predict modulus or toughness reduction over decades [31].

- Transfer learning allows models trained on one polymer system to adapt to another with minimal retraining, thus expanding applicability.
- Bayesian models provide not only predictions but also confidence intervals, which are essential for engineering design.

A case study is the prediction of modulus reduction in epoxy resins. By training a neural network on short-term creep tests, researchers were able to forecast the modulus decrease over 10^4 h, closely matching experimental data [32].

4.2 Degradation

Degradation in polymers is typically categorized as:

- **Thermal degradation**: Chain scission, crosslinking, or depolymerization caused by high temperatures.
- **Photo-degradation**: UV or visible light exposure leads to bond scission, oxidation, and discoloration.
- **Chemical degradation**: Reaction with solvents, acids, bases, or environmental pollutants.
- **Biodegradation**: Enzymatic or microbial breakdown in natural environments.

Predicting degradation requires understanding bond dissociation energies, reaction kinetics, and environmental interactions. Quantum chemical simulations provide insights but are restricted to small fragments and short timescales. Experimental degradation studies, while informative, cannot feasibly test every possible condition. AI offers several strategies:

- Regression models predict degradation rates from molecular descriptors and environmental conditions [13]. For example, decision trees or gradient boosting can relate polymer backbone chemistry, side groups, and UV exposure conditions to degradation half-lives.
- Classification models predict likely degradation pathways. For instance, given a polymer structure, an AI classifier can label it as prone to chain scission versus crosslinking.
- Hybrid QM + AI models accelerate prediction of bond scission pathways. A neural network coached on a limited set of DFT bond dissociation energies can predict which bonds in a polymer are most vulnerable under UV exposure [33].

In polycarbonate (PC), UV-induced degradation leads to yellowing and embrittlement. By combining experimental UV aging data with ML regression, researchers accurately predicted degradation rates under different light intensities and temperatures [34].

4.3 Creep and Relaxation Behavior

Creep refers to the time-dependent strain under constant stress, while relaxation describes the time-dependent stress under constant strain. These viscoelastic phenomena are critical in load-bearing polymer applications such as structural composites or biomedical implants. Traditional constitutive models, such as the Kelvin–Voigt or Maxwell models, use combinations of springs and dashpots to approximate viscoelasticity. While useful, these models often fail to capture nonlinearities at long timescales or under variable environmental conditions. AI models trained on experimental creep and relaxation datasets can learn to capture nonlinear time-dependent responses more effectively than analytical models. For example:

- Recurrent neural networks (RNNs) and long short-term memory (LSTM) networks are highly suited, as they naturally model sequential, time-dependent data [35].
- Hybrid approaches couple viscoelastic constitutive equations with AI correction terms, improving predictive accuracy without losing physical interpretability.
- Surrogate modeling replaces time-consuming finite element creep simulations with ML regressors, enabling rapid parametric studies.

In high-density polyethylene (HDPE), experimental creep data spanning thousands of hours were used to train an LSTM model. The AI predictions accurately extended creep curves to 10^5 h, outperforming traditional viscoelastic extrapolations [36].

5 Hybrid Approaches

While AI provides remarkable predictive capabilities, its black-box nature sometimes limits physical interpretability. Conversely, traditional polymer modeling methods such as molecular dynamics (MD), coarse-grained (CG) simulations, and continuum mechanics are grounded in physics but constrained by computational cost and scale. Hybrid approaches aim to combine the best of both: leveraging the accuracy of physics-based models with the speed and flexibility of AI. These synergistic frameworks are emerging as the most powerful strategy for multiscale polymer simulations, where AI acts as a bridge between simulation scales, or as an accelerator of computationally intensive tasks.

5.1 AI + Molecular Dynamics

Classical MD simulations resolve atomic motions with femtosecond timesteps, limiting simulations to nanoseconds–microseconds and system sizes up to millions of atoms. For polymers, whose relaxation times can span seconds to years, this poses a major challenge. Furthermore, force fields, the mathematical functions governing atomic interactions often require laborious parameterization and may not capture complex chemical reactivity or environmental conditions. AI enhances MD in several ways:

1. **Neural Network Force Fields (NNFFs)**

Instead of relying on predefined empirical force fields, AI models trained on quantum mechanical (QM) calculations can predict potential energy surfaces with near-DFT accuracy. Examples include DeepMD and ANI potentials, which significantly accelerate large-scale simulations [37].

2. **Surrogate Models for Long-Time Dynamics**

AI learns from short MD trajectories and extrapolates to predict longer timescale behaviors, bypassing brute-force simulation. This has been applied to predict polymer diffusion coefficients and glass transition behaviors [38].

3. **Accelerating Rare Event Sampling**

AI-guided enhanced sampling methods such as reinforcement learning identify and sample rare but critical transitions (e.g., chain scission or folding events).

In simulations of polyethylene crystallization, AI-driven NNFFs reproduced nucleation events at a fraction of the cost of QM-based MD, enabling accurate crystallization kinetics prediction [39].

5.2 *AI + Coarse-Grained Models*

Coarse graining reduces system complexity by representing groups of atoms (e.g., monomer units) as larger beads, enabling simulations over microseconds to milliseconds. However, deriving CG potentials that faithfully reproduce atomistic behavior is non-trivial. AI enhances CGM in many ways:

1. **Data-Driven CG Potential Derivation**

Instead of iterative Boltzmann inversion or force matching, neural networks can directly learn the mapping from atomistic simulations to CG force fields [40].

2. **Adaptive CG Models**

AI enables multiscale resolution by dynamically switching between atomistic and CG representations based on local chemical environments.

3. **Inverse Design**

AI trained on CG models can rapidly scan polymer design spaces (composition, chain length, functionalization) to predict structural and mechanical outcomes.

In a study on polystyrene melts, AI-learned CG models captured chain entanglements and viscosity predictions with higher fidelity than traditional CG potentials, while reducing computational cost by 50× [13].

5.3 *AI + Continuum Models*

Continuum-scale models (finite element methods, constitutive equations) are essential for predicting macroscopic polymer properties such as stress–strain behavior, fracture mechanics, and multiphase flow. However, constitutive models often rely on empirical fitting and cannot always capture molecular-level mechanisms. The hybrid strategies are as follows:

1. Physics-Informed Neural Networks (PINNs)

PINNs incorporate continuum-level governing equations (e.g., viscoelastic constitutive laws) into neural network training. This ensures predictions respect physical laws while improving accuracy [32].

2. AI as Constitutive Model Surrogates

Instead of predefined viscoelastic equations, AI can act as a "digital twin" of polymer stress-strain response, trained on experimental or MD/CG simulation datasets.

3. Bridging Micro to Macro

AI learns mapping functions from molecular/CG simulation outputs (e.g., relaxation spectra) to continuum-level constitutive parameters, effectively linking bottom-up simulation data with engineering-scale design models.

For polyurethane foams, a PINN-based hybrid model successfully predicted nonlinear stress–strain responses under varying strain rates, outperforming classical hyper elastic models [41].

6 Case Studies

The power of AI-integrated multiscale modeling is best appreciated through real-world case studies. These examples demonstrate how AI mediates between theoretical modeling and practical applications, enabling researchers to design new polymeric materials, predict long-term performance, and improve environmental sustainability.

6.1 *Case Study 1: Polymer Blend Miscibility Prediction*

Polymer blends are widely used to tailor mechanical, thermal, and chemical properties. However, miscibility prediction remains a challenge because it requires understanding molecular interactions across multiple scales. Traditional methods, such as Flory–Huggins theory, provide approximate guidance but often fail for complex, functionalized polymers. Recent studies have applied machine learning regression models trained on large experimental datasets of polymer blends. Features include Hansen solubility parameters, polarity indices, and polymer chain descriptors. By combining these with coarse-grained simulation outputs, ML models predict miscibility gaps and phase diagrams. Compared with thermodynamic models, ML approaches achieved significantly higher accuracy in predicting miscibility, particularly for biodegradable polymer blends (e.g., PLA/PCL, PLA/PBS). This accelerates the design of sustainable packaging and biomedical polymers [42].

6.2 Case Study 2: Long-Term Mechanical Degradation of PU Foams

Polyurethane (PU) foams are ubiquitous in automotive, aerospace, and biomedical applications. Their long-term mechanical durability is a key concern, especially under UV exposure and thermal cycling. Traditional accelerated aging tests are costly and time-consuming. Researchers combined accelerated UV-aging experiments with AI regression models to predict long-term modulus degradation. Neural networks trained on short-term exposure data were able to extrapolate modulus and compressive strength over decadal timescales. The hybrid model outperformed conventional empirical degradation laws, reducing prediction error by nearly 30%. This approach supports lifetime engineering of PU foams, reducing material waste and improving sustainability [43].

6.3 Case Study 3: UV-Induced Polymer Degradation Pathways

UV-induced degradation leads to chain scission, discoloration, and mechanical failure in polymers such as polycarbonate and polystyrene. Identifying degradation-prone bonds requires extensive DFT simulations of bond dissociation energies, which are computationally prohibitive for large systems. A hybrid DFT + ML framework was developed, where quantum-level data on small molecular fragments were used to train AI models capable of predicting bond scission probabilities in larger polymers. This approach captured photo-oxidation pathways and identified the most vulnerable sites within polymer backbones. The AI-enhanced workflow enabled high-throughput screening of polymer stability, guiding the development of UV-stabilized formulations for outdoor coatings, solar panel encapsulants, and medical devices [44].

6.4 Case Study 4: Data-Driven Prediction of Glass Transition Temperature (T_g)

The glass transition temperature is a core property that governs polymer usability in applications ranging from packaging to aerospace. Traditional T_g prediction requires exhaustive DSC (differential scanning calorimetry) experiments or long-timescale simulations. ML regression models were trained on large databases containing polymer structures, descriptors (molecular weight, polarity, backbone rigidity), and experimental T_g values. Random forest and gradient boosting algorithms emerged as the most accurate predictors. These models not only predicted T_g with high accuracy but also provided feature importance rankings (e.g., aromaticity and hydrogen bonding capacity strongly influence T_g). This helps researchers design high-performance thermoplastics without trial-and-error experimentation [45].

6.5 Case Study 5: Biodegradation Rate Prediction for Sustainable Polymers

With the increasing demand for eco-friendly materials, predicting biodegradation rates of polymers (e.g., PLA, PHA, starch blends) is essential. Experimental biodegradation studies are slow and highly dependent on environmental conditions. By training supervised ML models on datasets of biodegradation under soil, compost, and aquatic conditions, researchers developed predictive models linking polymer chemistry and microstructure to degradation rates. This approach provides a rapid screening tool for designing polymers with tailored degradation profiles, supporting the development of next-generation bioplastics [46].

7 Challenges and Future Perspectives

The integration of AI with multiscale polymer modeling has brought about significant advances, yet several challenges remain before it can become a universally adopted framework for research and industrial applications. These challenges span issues of data quality, interpretability, computational efficiency, and integration with experimental pipelines. Addressing them will shape the trajectory of AI-driven polymer science in the coming decades.

7.1 *Data Quality and Availability*

High-quality datasets are the foundation of reliable AI models. In polymer science, however, data availability is limited by:

- **Heterogeneity of experimental conditions**—Properties such as T_g or modulus vary with processing history, making it difficult to create standardized datasets.
- **Scarcity of open databases**—Unlike small molecules, polymers lack large, standardized repositories.
- **High cost of QM and MD simulations**—Limits the scale of synthetic training data.

Potential Solutions

- Data augmentation using generative models (GANs, variational autoencoders) to create synthetic polymer structures and property datasets.
- Transfer learning, where AI models pretrained on large chemistry datasets are fine-tuned for polymers.
- Collaborative data platforms (e.g., Materials Project, Polymer Genome) that encourage open data sharing [47].

7.2 *Interpretability of AI Models*

Deep learning models are often widely challenged as black boxes. While they may predict properties with high accuracy, they do not always reveal the underlying structure–property relationships. This reduces their usefulness for scientific discovery.

Potential Solutions

- Explainable AI (XAI) techniques such as SHAP (SHapley Additive exPlanations) and LIME (Local Interpretable Model-Agnostic Explanations) provide micro-level insights.
- Hybrid physics-informed AI models that embed physical constraints (e.g., conservation laws, thermodynamic limits) into neural networks, ensuring outputs remain scientifically meaningful [48].
- Model simplification and visualization tools for polymer researchers to explore the influence of structural features.

7.3 *Computational Cost Versus Accuracy*

Although AI accelerates simulations, there remains a trade-off between accuracy and computational feasibility:

- DFT/QM-based training sets provide high accuracy but are prohibitively expensive.
- Classical MD/CG data are cheap but may sacrifice fidelity.
- Training large neural networks requires substantial computational infrastructure (GPUs, HPC).

Potential Solutions

- Active learning frameworks that selectively query new simulations to optimize dataset efficiency.
- Multi fidelity modeling, where AI integrates low-accuracy, high-volume data with sparse high-accuracy data.
- Federated learning approaches to distribute training workloads across institutions without sharing proprietary data.

7.4 Integration with Experiments

AI-driven models are often validated against simulations rather than experiments, leading to a simulation-to-reality gap. Additionally, experimental reproducibility issues limit AI's reliability.

Potential Solutions

- Self-driving laboratories that integrate automated synthesis, characterization, and AI-driven optimization.
- Closed-loop feedback systems where AI models guide experiments, and experimental results refine the AI in real time.
- Examples include AI-controlled polymerization reactors that dynamically adjust conditions to achieve target molecular weights or mechanical properties [49].

7.5 Future Trends

1. AI-driven pipelines will autonomously design, synthesize, and test new polymers with minimal human intervention.
2. RL agents will optimize polymer synthesis routes or processing conditions based on reward functions (e.g., maximizing toughness, minimizing cost).
3. AI will help design biodegradable polymers, predict life cycle impacts, and optimize recycling processes, supporting circular economy initiatives.
4. Real-time virtual models of polymer systems, continuously updated with experimental/operational data, will enable predictive maintenance and performance optimization in industrial applications.
5. Advances in quantum computing could dramatically reduce the cost of QM simulations, enabling richer training datasets for AI models [50].

8 Conclusion

The field of polymer science is shifting through a phase of change driven by the convergence of multiscale modeling and artificial intelligence (AI). Traditional approaches ranging from quantum mechanical (QM) calculations to molecular dynamics (MD), coarse-grained (CG) simulations, and continuum mechanics (FEM) have provided invaluable insights into polymer behavior. However, their inherent scale and time limitations restrict the scope of problems they can address. AI, with its ability to learn patterns, extrapolate behaviors, and integrate heterogeneous datasets, provides a powerful complementary tool that extends the reach of conventional methods. Throughout this chapter, we examined how AI augments polymer modeling at multiple scales:

- Atomistic scale: AI-driven force fields achieve near-DFT accuracy while enabling large-scale simulations, bridging the gap between precision and efficiency.
- Mesoscopic scale: Data-driven coarse-graining leverages ML models to derive effective potentials, accelerating simulations of complex polymer assemblies.
- Macroscopic scale: Physics-informed AI models enhance continuum mechanics, enabling predictive stress–strain and failure modeling of large polymeric structures.
- Long-term phenomena: Neural networks and regression models predict aging, degradation, creep, and relaxation behavior far beyond the temporal limits of MD or experiments.

By integrating AI into each of these scales, researchers are now able to link atomistic details with macroscopic performance, offering a holistic predictive framework that was previously unattainable. The most promising direction lies in hybrid approaches, where AI does not replace physics-based modeling but rather accelerates, augments, and interprets it. Neural network force fields, AI-assisted coarse-graining, and physics-informed neural networks exemplify this synergy. Such methods not only increase computational efficiency but also preserve scientific interpretability, ensuring that predictions remain grounded in physical reality. Looking ahead, the integration of AI with multiscale polymer modeling holds the potential to transform materials science into a predictive, autonomous, and sustainable discipline. Some emerging directions include:

- Autonomous materials discovery platforms that design, synthesize, and test polymers with minimal human intervention.
- Reinforcement learning frameworks for optimizing polymer synthesis and processing conditions.
- Sustainability-driven AI applications, guiding the design of biodegradable polymers, recycling pathways, and low-carbon materials.
- Digital twins of polymeric systems for real-time monitoring, predictive maintenance, and industrial optimization.

In essence, the convergence of AI and multiscale modeling is more than a technological advancement it represents a paradigm shift. By uniting data-driven intelligence with physics-based rigor, researchers can accelerate polymer innovation, reduce trial-and-error experimentation, and contribute to solving global challenges in energy, environment, and healthcare.

References

1. Butler, K.T., Davies, D.W., Cartwright, H., Isayev, O., Walsh, A.: Machine learning for molecular and materials science. Nature **559**(7715), 547–555 (2018)
2. Carleo, G., Cirac, J.I., Cranmer, K., Daudet, L., Schuld, M., Tishby, N., et al.: Machine learning and the physical sciences. Rev. Mod. Phys. **91**(4), 045002 (2019)
3. Chen, C., Zuo, Y., Ye, W., Li, X., Deng, Z., Ong, S.P., Luo, J.: A critical review of machine learning of energy materials. Adv. Energy Mater. **11**(8), 2002740 (2021)
4. Faller, R., Müller-Plathe, F.: Chain stiffness intensifies the microphase separation in binary polymer blends: a molecular dynamics study. Macromolecules **34**(4), 1436–1443 (2001)
5. Karimi-Varzaneh, H.A., Müller-Plathe, F.: Coarse-grained modeling for polymers. In: Advances in Polymer Science, vol. 252, pp. 295–321. Springer (2012)
6. Li, Y., Li, H., Zhang, J., Liu, J.: Multiscale simulation of mechanical properties of polymer nanocomposites. Compos. B Eng. **85**, 28–35 (2016)
7. Liu, X., Zhang, H., Müller-Plathe, F.: Multiscale modeling of polymers: a review of recent developments. Polymer **207**, 122929 (2020)
8. Müller-Plathe, F.: Coarse-graining in polymer simulation: from the atomistic to the mesoscopic scale and back. ChemPhysChem **3**(9), 754–769 (2002)
9. Peter, C., Kremer, K.: Multiscale simulation of soft matter systems—from the atomistic to the coarse-grained level and back. Soft Matter **5**(22), 4357–4366 (2009)
10. Ramprasad, R., Batra, R., Pilania, G., Mannodi-Kanakkithodi, A., Kim, C.: Machine learning in materials informatics: recent applications and prospects. npj Comput. Mater. **3**(1), 1–13 (2017)
11. Schmidt, J., Marques, M.R.G., Botti, S., Marques, M.A.L.: Recent advances and applications of machine learning in solid-state materials science. npj Comput. Mater. **5**(1), 1–36 (2019)
12. Tervoort, T.A., Govaert, L.E.: Time-dependent mechanical behavior of polymers: a review. Prog. Polym. Sci. **113**, 101343 (2021)
13. Zhang, L., Han, J., Wang, H., Car, R., Weinan, E.: Deep potential molecular dynamics: a scalable model with the accuracy of quantum mechanics. Phys. Rev. Lett. **120**(14), 143001 (2018)
14. Kohn, W., Sham, L.J.: Self-consistent equations including exchange and correlation effects. Phys. Rev. **140**(4A), A1133–A1138 (1965)
15. Parr, R.G., Yang, W.: Density-Functional Theory of Atoms and Molecules. Oxford University Press (1989)
16. Allen, M.P., Tildesley, D.J.: Computer Simulation of Liquids, 2nd edn. Oxford University Press (2017)
17. van Duin, A.C.T., Dasgupta, S., Lorant, F., Goddard, W.A.: ReaxFF: a reactive force field for hydrocarbons. J. Phys. Chem. A **105**(41), 9396–9409 (2001)
18. Marrink, S.J., Risselada, H.J., Yefimov, S., Tieleman, D.P., de Vries, A.H.: The MARTINI force field: coarse grained model for biomolecular simulations. J. Phys. Chem. B **111**(27), 7812–7824 (2007)
19. Hoogerbrugge, P.J., Koelman, J.M.V.A.: Simulating microscopic hydrodynamic phenomena with dissipative particle dynamics. Europhys. Lett. **19**(3), 155–160 (1992)

20. Español, P., Warren, P.: Statistical mechanics of dissipative particle dynamics. Europhys. Lett. **30**(4), 191–196 (1995)
21. Andersons, J., König, M., Hojo, M., Ochiai, S.: Stress intensity factors for a crack in a viscoelastic polymer. Eng. Fract. Mech. **72**(6), 887–897 (2005)
22. Bergström, J.S., Boyce, M.C.: Constitutive modeling of the large strain time-dependent behavior of elastomers. J. Mech. Phys. Solids **46**(5), 931–954 (1998)
23. Liu, P., Gao, X., Ma, J., Xu, J.: Multiscale modeling of polymer/clay nanocomposites. Compos. Sci. Technol. **69**(11–12), 2188–2194 (2009)
24. Calvert, P.: Hydrogels for soft machines. Adv. Mater. **21**(7), 743–756 (2009)
25. Wang, Z.G., Larson, R.G., Lodge, T.P.: Phase behavior of polymer blends from multiscale simulations. Macromolecules **42**(21), 7953–7963 (2009)
26. Kim, C., Chandrasekaran, A., Huan, T.D., Das, D., Ramprasad, R.: Polymer informatics: current status and critical next steps. J. Phys. Chem. C **122**(31), 17575–17585 (2018)
27. Wu, S., Kondo, Y., Kakimoto, M., Yang, B., Yamada, H., Kuwajima, I., Lambard, G., Hongo, K., Xu, Y., Shiomi, J., Yoshida, R.: Machine-learning-assisted discovery of polymers with high thermal conductivity using a molecular design algorithm. npj Comput. Mater. **5**, 66 (2019)
28. Zhao, Z., Zhang, W.: Machine learning in polymer informatics. Polymer **179**, 121628 (2019)
29. Popova, M., Isayev, O., Tropsha, A.: Deep reinforcement learning for de novo drug design. Sci. Adv. **4**(7), eaap7885 (2018) (Concept transferable to polymer synthesis optimization)
30. Smith, J.S., Isayev, O., Roitberg, A.E.: ANI-1: an extensible neural network potential with DFT accuracy at force field computational cost. Chem. Sci. **8**(4), 3192–3203 (2017)
31. Wang, H., Zhang, L., Han, J., Weinan, E.: DeePMD-kit: a deep learning package for many-body potential energy representation and molecular dynamics. Comput. Phys. Commun. **228**, 178–184 (2018)
32. Wang, J., Olsson, S., Wehmeyer, C., Pérez, A., Charron, N.E., de Fabritiis, G., Noé, F., Clementi, C.: Machine learning of coarse-grained molecular dynamics force fields. ACS Cent. Sci. **5**(5), 755–767 (2019)
33. Struik, L.C.E.: Physical Aging in Amorphous Polymers and Other Materials. Elsevier (1978)
34. Chen, L., Pilania, G., Batra, R., Huan, T.D., Kim, C., Kuenneth, C., Ramprasad, R.: Polymer informatics: Current status and critical next steps. Prog. Polym. Sci. **117**, 101724 (2021)
35. Klompen, E.T.J., Engels, T.A.P., Govaert, L.E., Meijer, H.E.H.: Modeling of the postyield response of glassy polymers: influence of physical aging. Macromolecules **38**(16), 6997–7008 (2005)
36. Myung, J., Kim, S., Park, J., Lee, T.S.: Machine learning for polymer degradation prediction. Polym. Degrad. Stab. **176**, 109149 (2020)
37. Qiao, Z., Olsson, S., Wehmeyer, C., Noé, F., Clementi, C., Huang, X.: Data-driven coarse-graining for polymer degradation pathways. J. Chem. Theory Comput. **16**(4), 2535–2546 (2020)
38. Decker, C., Bendaikha, T.: Photo-oxidation of polycarbonate. Polym. Degrad. Stab. **60**(3), 341–353 (1998)
39. Yang, Z., Yu, Y., Xu, X.: Long short-term memory networks for creep prediction of polymers. Mater. Today Commun. **21**, 100695 (2019)
40. Zhang, X., Li, H., Huang, Y.: Machine learning prediction of creep behavior in polyethylene. Polym. Test. **93**, 106912 (2021)
41. Behler, J.: Perspective: machine learning potentials for atomistic simulations. J. Chem. Phys. **145**(17), 170901 (2016)
42. Husic, B.E., Noé, F.: Markov state models: from an art to a science. J. Am. Chem. Soc. **140**(7), 2386–2396 (2018)
43. Wang, H., Gómez-Bombarelli, R., Kremer, K.: Machine learning coarse-grained models for polymers. Macromolecules **53**(22), 9449–9460 (2020)
44. Raissi, M., Perdikaris, P., Karniadakis, G.E.: Physics-informed neural networks: a deep learning framework for solving forward and inverse problems. J. Comput. Phys. **378**, 686–707 (2019)
45. Liu, Y., Liu, Z., Zhang, J., Yu, Y.: Hybrid machine learning and constitutive modeling of polyurethane foams. Compos. Sci. Technol. **207**, 108725 (2021)

46. Ma, R., Wu, J., Gao, H., Zhang, G., Yao, Y.: Machine learning-enabled prediction of polymer blend miscibility. Polymer **220**, 123563 (2021)
47. Liu, Z., Chen, H., Zhang, Y.: Neural network modeling of polyurethane foam degradation under UV exposure. Polym. Degrad. Stab. **181**, 109360 (2020)
48. Li, J., Wang, H., Xu, D.: Hybrid machine learning and quantum simulations for predicting polymer degradation pathways. Macromolecules **52**(24), 9508–9518 (2019)
49. Chen, L., Pilania, G., Batra, R.: Machine learning models for glass transition temperature of polymers. J. Chem. Inf. Model. **61**(3), 1345–1356 (2021)
50. Khetan, A., Gurumurthy, A., Maranas, J.K.: Data-driven modeling of polymer biodegradation for sustainable materials design. Green Chem. **21**(15), 4126–4138 (2019)

Data-Driven Polymer Informatics

Deepika Goswami, Anzala Mumtaz, Hema Arya, Kiran Negi, and Sandeep Dutt Maindoli

Abstract The domain of materials science and engineering, especially polymer science is beholding a notable transition in the rate of discovery, predominantly influenced by the integration of data-driven informatics approaches, which denotes a growth towards more anticipatory, efficient, and interdisciplinary research. Polymers are highly complex and versatile structures. Due to the wide-ranging attributes and versatility, polymeric materials are extensively used worldwide in numerous areas like medicine, consumer goods, aerospace, packaging of plastics, pot and pan handles, soft drinks bottles, and electronics. Classical approaches for synthesizing polymers are usually time-consuming and resource-demanding; Nevertheless, data-driven approaches facilitate the speedy exploration of large-scale chemical and structural domains. The advent of data-driven approaches has led to the rise of a new domain called "Polymer Informatics", which utilizes Artificial Intelligence (AI) and Machine Learning (ML) for the discovery and development of new polymers. In recent years, machine learning has garnered significant attention and has become a powerful tool in many research fields. Machine learning contributes to resolving the time-consuming trial-and-error technique in conventional methods. Despite notable development, polymer informatics faces challenges that restrain its growth towards durable material discovery. Therefore, the data-driven practices are more effective and workable than the time-consuming experimental procedures. Some successful attempts in polymer informatics have been made, such as for quick polymer prediction, surrogate models are trained on pre-existing polymer data, facilitating the screening of promising polymer candidates with definite target property requirements. Several tools and software programs have been developed and designed to foster the incorporation of machine learning in polymer science, which are available to assist data management, analysis, and modeling. Open source tools and platforms are also designed to benefit the researchers in synthesizing a novel polymer without any hindrance.

Keywords Polymer informatics · Data-driven approaches · Machine learning · Deep learning

---

D. Goswami · A. Mumtaz (✉) · H. Arya · K. Negi · S. D. Maindoli
Department of Zoology, Kumaun University, Nainital, Uttarakhand, India
e-mail: anzalamumtaz122@gmail.com

B. Sharma et al. (eds.), *AI-Driven Polymer Design*, Engineering Materials,
https://doi.org/10.1007/978-981-95-8353-9_5

1 Introduction

In recent years, the field of materials science has experienced a rapid transformation, fueled by the explosive growth of data and advancements in computational technologies. To accelerate polymer research and development, researchers are increasingly turning to machine learning and data analysis instead of relying solely on lengthy and costly experiments [38]. The term "polymer" originates from the Greek words "polus," meaning "many," and "meros," meaning "parts." A polymer is a large molecule, also known as a macromolecule, that consists of smaller repeating units called monomers, which are connected by covalent bonds. Polymers display a wide variety of parameters, incorporating mechanical, thermal, optical, electrical, magnetic, and biological attributes like convenience of processing, stability, durability, and environmental sustainability [36]. There are two main types of polymers: natural polymers, such as DNA, proteins, and cellulose, etc. and synthetic polymers, like nylon, polyester, and polyethylene. Polymers serve as the foundation for a vast array of products, ranging from everyday plastics to advanced biomedical devices. In the past, traditional methods for discovering new polymers and optimizing their properties relied on trial-and-error experiments and expert intuition. However, these methods were often time-consuming and expensive [37].

The advent of data-driven approaches has led to the emergence of a new field called polymer informatics. This interdisciplinary area combines materials science, chemistry, and data science to predict and understand polymer properties using data-driven techniques. Polymer informatics is a field that utilizes tools like AI and Machine Learning for the discovery and development of polymers. For instant polymer prediction, surrogate models are trained on existing polymer data which enables screening of favorable polymer candidates with definite target property requirements. In recent years, Polymer Informatics has exhibited substantial success in predicting polymer properties by incorporating data-driven models to create structure–property relationships [6, 15].

Ordinarily, machine learning models attain this by linking molecular features or fingerprints of discrete monomers to their equivalent polymer properties. Instead of relying solely on manually made features, graph neural networks are the best alternative, especially for datasets ranging from moderate to large size with over 300 samples, by autonomously detecting structural elements relevant to a specific property [6, 11].

Consequently, machine learning models may face a challenge with accuracy when the appropriate chemical space is underacknowledged in the training data. Integrating machine learning with physics-based modeling facilitates a collaborative approach. Machine learning promotes the rapid screening of polymer candidates, whereas physics-driven simulations present more in-depth validation of promising options. This coordinated strategy can amplify the possibility of valid experimental results as opposed to using either method separately [6].

However, Artificial Intelligence (AI) techniques have a great impact on human life, science, and technology as it is time-efficient and less expensive than traditional methods. In this field of research, old methods like trial-and-error face many obstacles due to the extensive combinatorial complexity. Therefore, Machine Learning (ML) is opening new doors for the discovery of novel polymers by predicting material properties, constructing new polymer designs, and enriching synthesis conditions [7]. With the support of high computing power and advanced algorithms, Machine Learning, a subfield of Artificial Intelligence (AI), has advanced swiftly in the last several years and is proficient in absorbing and executing high-dimensional data [27]. The astonishing chemical and physical diversity of polymers, both at small and large scales, is being studied by using Data-driven strategies in order to deal with significant challenges [5].

In addition to these challenges, some major obstacles for polymer informatics are that carefully assembled and structured data are limited, as well as strategies for generating machine-readable formats that focus both on the complex structures of polymers and synthesis conditions [5]. Thus, to address these major obstacles, "Polymer Informatics" plays an important role.

This chapter provides a holistic overview of data-driven polymer informatics. We begin the chapter by discussing the key elements of polymer science and data acquisition and generation, need for polymer informatics. We then explore different machine learning methods/models and strategies to develop polymers.

2 Essential Elements of Polymer Informatics

Polymer Informatics is an interdisciplinary field that uses machine learning and AI tools to discover new polymers. There are five main elements of Polymer Informatics.

One of the core constituents of polymer informatics is the collection of the data from experiments and high-throughput computations. Contrary to hard materials, availability of limited well-organized or clean polymer data is found to be present which can be used for machine learning or artificial intelligence (AI) based techniques.

Data Generation and Acquisition:

If an adequate volume of polymer data can be significantly generated or curated, it can lead to the discovery and design of functional polymers with strategic performance. Below we will discuss how polymer data can be aggregated from the "scientific literature" or synthesized using "high-throughput" and "autonomous computations".

Scientific Literature

Scientific literature is one of the helpful data resources which is steady and vast. It includes—Printed handbooks, Journal Articles, Online Repositories [21]. Of late, for easy access to polymer data, numerous polymer databases have been digitalized. Some representative databases are listed below: PoLyInfo, CROW Polymer Property Database, Polymers: A Property Database, CAMPUS, LANDOLT-BORNSTEIN, Polymer Property Predictor and Database (NIST) [5]. However, large volumes of experimental data remain stuck in the scientific literature in polymer handbooks and online-repositories which is sporadically extracted through time-consuming manual effort.

Alternate Approach: NLP (Natural Language Processing) could be used to extract polymer data continuously. Another significant resource of polymer data is high-throughput computations using DFT (Density Functional Theory) and Molecular Dynamics (MD) simulations [5].

High-Throughput and Autonomous Computational Agents:

For developing and screening polymer data like structures, polymer properties and its processing conditions, high-throughput or automated large-scale methods are required. Two important approaches can be used to organize polymer data which are First-principles theory and Classical MD Simulations. Due to the complex chemical and physical structure of polymers, it is not easy to compile the data. In general, polymers exhibit either amorphous or semi-crystalline phase. To emulate polymers and evaluate their physical, electronic, and dielectric properties, researchers have formulated small-scale models (less than 100 atoms) due to the expensive computational cost of first-principles calculations [12, 13].

However, the estimated result is affected by the accuracy issues depending on the methodology used. To imitate polymers at large-scales, classical molecular dynamics (MD) using empirical force fields have been implemented to explore their structural dynamics and various properties like dielectric, thermal, mechanical and ion transport [1, 5, 20, 41].

Descriptors and Polymer Representation: Second important component of polymer informatics is an appropriate framework to develop machine-readable polymer representations. However, transforming polymers into numerical fingerprints and making it modifiable to machine learning models. These numerical representations are called as Descriptors, Features or Fingerprints. Some polymer representation methods are- SMILES (Simplified Molecular Input Line Entry System) string representation, this is considered to be one of the popular descriptors for macromolecules.

The original SMILES (Simplified Molecular Input Line Entry System) format was proposed in 1988 and it was widely accepted in chemical community. This string representations are mostly used for the chemical space exploration of polymers. A SMILES string representing a restricted chemical space can be altered into a two-dimensional binary matrix which can also be used as input for a machine learning model to predict polymer properties. Furthermore, BigSMILES (an extension to SMILES), has been developed lately. The main purpose of BigSMILES is describing macromolecules like homo- and co-polymers. However, recently there are no actual fingerprinting algorithm proficient for transforming BigSMILES into numerical vectors [17].

Polymer Property Prediction and Design: The third element is constructing models for prediction of polymer property and its design. To select appropriate learning procedures for mapping fingerprints and properties of polymer is a foundational step. Numerous learning procedures have been implemented depending on the following measures like complexity of the destination property, the nature and volume of the existing data collections.

3 Steps Involved in Machine Learning-Driven Polymer Innovation

Step 1: Collection of Datasets: The first and foremost step in designing a polymer is the collection of the datasets. There are four types of approaches to collect the data points which are as follows- Collection of datasets from existing database, Collection of datasets from literatures, Collection of datasets from scientific calculations or conceptual models, and Collection of datasets from experiments.

Step 2: Fingerprinting: In this step, conversion of 3D molecules into tensors or scalar value is done. There are different types of fingerprinting methods discussed below

Linear Notation Encoding and Direct Label Representation: There are two procedures in this method-Linear notation and Number encoding.

Linear Notation: In this procedure, the 3D molecule is converted into a linear sequence of characters or symbols. Widely used format of linear notations involve are SMILES (Simplified Molecular Input Line Entry System), WLN (Wiswesser Line Notation), MCDL (Modular Chemical Descriptor Language) [8, 38], and BigSMILES.

Number Encoding: In this procedure, linear string is converted into numbers. Some common methods used are one-hot and ASCII conversion.

One-hot Conversion In one-hot conversion, to implement one-hot encoding, a dictionary defining the encoding scheme must be provided. The dictionary can be obtained by compiling all the atom types, bonds, numerical values, and other symbols present in the training dataset, expressed in combination as a text string. **Example:** To convert 2,5-difluorostyrene into a binary matrix, there are three steps which must be followed which are- To start, the SMILES code is written in the first column, whereas the dictionary is in the first row. Keeping in mind that some cheminformatics tools and few websites can convert a molecule of polymer into its SMILES representation, for instance, this can be done utilizing the RDKit package or via the "Cheminfo" website. Second step is, if the overlapping points represent the same character, a "1" is entered into the matrix. Then the last step is that, a "0" is written in for all remaining positions in the matrix.

Morgan Fingerprinting: With the help of this method, we can acquire a rare, sequential numbering of atoms for any given monomer. This technique was invented by

Rogers and Hahn. The Morgan fingerprinting method is a modified version of the Extended Connectivity Fingerprint (ECFP) [26, 38]. There are mainly three procedures involved in this method. In the first procedure, each atom in the monomer is appointed an integer. In the second procedure, each atom identifiers are updated to demonstrate each atom's neighbor. Finally, in the last procedure, the same features are discarded to construct a single representative fingerprint.

Integrated Tensor Representation Using Compositional Blocks: This method comprises of a tensor combination based on distinct repetitive blocks in a polymer.

4 Case Study

A study conducted by Mannodi-Kanakkithodi et al. [19] demonstrated that the polymers incorporates seven distinct building blocks and three different surrounding blocks, therefore they created three types of tensors, featuring &-dimensional vector, 7×7 matrix, and $7 \times 7 \times 7$ matrix, to depict the building blocks, neighboring blocks, and triplet neighboring blocks, respectively.

Molecular Graph: In this technique, atoms and chemical bonds are depicted as vertices and edges, enabling the whole monomer to be illustrated as an undirected graph. It is significant to mention that this illustration is precisely appropriate for use with Graph Convolutional Neural Networks (GCNNs) [32].

Quantitative Structure–Property Relationship (QSPRs) Descriptors: There are two main steps involved in this technique-First step is the Detection of all parameters that may be associated with the prediction objective. The second step is enforcing scientific software or tools to evaluate these parameters.

k-mer Method: This is the standard approach used in bioinformatics which offers to efficiently represent the sequence for DNAs. This method seeks to display the different neighboring relations in a polymer chain.

Step 3: Machine Learning Framework: In this step, a variety of models were utlilized by researchers to investigate the relationship between the polymer fingerprints and polymer properties. Several traditional machine learning models are discussed below-

Common learning procedures implemented are as follows:

Linear or Non-linear Regression Algorithms: According to the linear regression algorithm, it is presumed that the polymer property which is being constructed is in actual a linear hypothesis function of the fingerprints, therefore, this is the easiest procedure to construct machine learning models. Several polymer prediction models have been engineered via group contribution methods, multiple linear regression, and least-squares regression, QSPR or quantum-chemical fingerprints.

However, these models are restricted by the overlook of the non-linear relationship between polymer fingerprints and their properties. For that reason, non-linear regression algorithms have been enforced to combat this problem. Some examples are Support Vector Machine (SVM), Kernel Ridge egression (KRR), and Gaussian Process Regression (GPR) [5, 34].

Multi-fidelity Information Fusion Strategies:

It is natural to encounter challenges encompassing multiple data collections that are different in terms of accuracy, creation cost, and noise levels. Typically, the most precise experimental results or algorithmic outputs which is known as High-Fidelity (HF) Data. These HF data are higly expensive and labor-intensive to acquire [5]. Nevertheless, this could be achieved through less-expensive methods at lower accuracy. Properties of the polymer of interest can be roughly calculated with further cost-effective techniques, though that may be at lower accuracy. Some techniques like Empirical correlations, basic group contribution method, and also less-intensive Quantum mechanical simulations can offer what we now called as Low-fidelity (LF) data. In such situations, to accomplish highly precised and trustworthy property predictions feasible at the high-fidelity level, a "Multi-Fidelity (MF) information fusion model strives to merge data from these different sources of inconsistent quality [16]. Studies comparing different techniques have exhibited that the multi-fidelity methods attain higher prediction accuracy than any single-fidelity methods especially when working with small high-fidelity databases.

Deep Neural Networks:

Deep Neural Networks (NN) are being progressively applies in Polymer Informatics. Traditional machine learning methods offer good property prediction accuracy, yet these methods are programmatically effective for only systems with small datasets.

Case Study:

For solvent prediction, scientists have developed a deep neural network model. An important step in synthesizing polymer is choosing an appropriate polymer–solvent pairs. In this model, the scientists trained a binary classification NN model, where the model anticipates whether it is a good solvent or no-solvent for a given polymer–solvent pair. Here 4,595 polymers and 24 solvents, constituting a total of 11,958 polymers along with good solvent pairs and 8,469 polymers together with non-solvent pairs were used. A highly designed multilayer perceptron (a simple type of artificial neural network model) was leveraged, demonstrating a innovative architecture [23]. There were two sections in neural network. The first section comprises of two separate input branches, of the two branches one prepared polymer descriptors which was derived from hierarchial fragment-based fingerprint technique. While the other branch dealt with solvent descriptors depicted by "One-Hot Encoding" technique. In the following stages, the latent representation of both the polymers and the solvent were mixed into a single merged latent vector [4].

5 Designing a Polymer

Purpose: The main purpose of designing a polymer is to develop materials which comprises of a unique and desirable properties customized for specific applications in real world.

Polymer Design-In order to design a polymer, First-Principle Methods or Predictive Machine Learning Models or even their amalgamation can be used. Traditionally, there are three main steps involved in designing a polymer which are- encompassing a large chemical space with adequate candidates grounded in on field-specific knowledge. Next, integrating all the feasible candidates with representations, and lastly, by contrasting and screening candidates with targeted polymer design properties [38]. In the past, all the data-driven approaches used for testing polymer configurational space and its design can be organized into four classes:

Edisonian Design: To evaluate the property of a large number of candidate material, high-throughput computations or experiments are used. For in-depth analysis and potential use in a particular application. The small number of top candidates are recognized from the datasets for more investigation and potential use in a particular application. This method follows conventional Trial-and-Error Edisonian Technique. Here, the structure of the candidates are shortlisted either spontaneously or based on an intuitive understanding of a polymer's configurational and chemical characteristics [22].

Machine Learning (ML) Evaluated Design: For constructing an ML Model, a pre-established structure property dataset is employed. For predicting the large number of candidate material, ML Model is applied. The leading candidates extrapolated by ML Model are selected for further investigation using First-Principles Method and additional assessment.

First-Principle-Method Based Inverse Design: In order to achieve a desired polymer property, inverse design is one of the best methods for identifying most effective set of parameters for that property. This technique combines an optimization algorithm with first-principle-based computational or experimental estimation of the candidate polymers. The optimization algorithm repeatedly formulate new candidates, whose properties are evaluated through experiments.

ML-Based Inverse Design: Since First-Principle-Based Inverse Design faces a challenge of real-time evaluation of polymer properties during each design cycle (a time-consuming step), to overcome this challenge, recent techniques have shifted to Machine Learning (ML) Models to predict polymer properties rather than measuring them directly (Table 1).

Step4: New Polymer Generation:

In this process, machine learning frameworks are employed to construct novel chemical structures of polymers. Based on the type of polymers, these frameworks are

Table 1 Comparison between ML models and first principle methods

S. no	ML models	First principle methods
1	ML models are used to predict the property of unknown polymers	First principle methods are used for the inverse design of polymers for identification of new polymers
2	Generate low-fidelity because they can give estimation of materials properties when the test data are within the range of their training data	Generate high-fidelity data to train machine learning ML Model when experimental data is limited
3	Methods involved: Supervised learning; unsupervised learning; deep learning models	Method involved: DFT (density functional theory), molecular orbital theory, quantum molecular dynamics

classified into two categories: First is synthesizing new homopolymers and the second is synthesizing new copolymers.

New Polymer Synthesis for Homopolymers: Homopolymers are the polymers with one repeat unit [30]. The synthesis of homopolymer is done by sinlge-unit cell-based representations of chemical structures (monomer). There are four traditional methods for synthesizing chemical structures of homopolymers. It is important to highlight that these approaches are also used in case of drug molecules because drugs are usually synthesized from small molecules.**Variational Autoencoder (VAE), Generative Adversarial Network (GAN), Genetic Algorithm (GA), and Monte Carlo Algorithm (MC)**.

New Polymer Synthesis for Copolymers: Copolymers are the polymers with more than one kind of monomer. In the domain of machine learning-enabled polymer research, the majority of papers focuses on homopolymers, as of now, few papers have conducted studies in this particular domain for copolymers [39, 40]. A method called Monomer Combination Method (MCM) was introduced. This approach can be broken down into two stages: First, they gathered monomers and crosslinkers from available literature. Next, they created various combinations of these monomers and crosslinkers, reinforcing that each arrangement conformed to established chemical reaction rules. To give an example, the monomers featuring C=C bonds are unable to undergo reactions with epoxides and sould not be paired in the latter rearrangement. By implementing this strategy, they uncovered 14 new TSMPs (Thermoset Shape Memory Polymers, representing strengthened recovery stress [40] and 5 novel UV-curable TSMPs comprising desirable characteristics [39].

Online User-Interface: For convenient and rapid public access to the developed metamodels, online user-interface is the best option.

AI-based Synthesis: Last element of polymer informatics is the artificial intelligence based polymer synthesis.

6 Need for Informatics in Polymer Research

Since polymers are versatile and used in everyday life, from plastic to aerospace and biomedical devices etc. However, it is challenging to develop polymers especially with traditional methods (trial-and-error and expert's intuition) because these are time-consuming which often takes 15–25 years. In this fast-paced life, it is very important to resolve this issue, for that Polymer Informatics comes into play.

Thus, actions are initiated to draw attention to the fact that how data-driven strategies help in constituting emerging links and fostering the fundamental knowledge of polymer science [22].

7 The Role of Data in Polymer Research

In Polymer Informatics, the databases play a pivotal role. In recent years, numerous polymer-related databases are being constructed and improved. For instance, PoLyInfo (it is a polymeric database which serves as a resource for researchers in polymeric material design). By October 2023, PoLyInfo maintained records of data on 18,697 homopolymers, 7737 copolymers, and had 494,820 property points from 161,464 polymer samples. PoLyInfo database had covered approximately 1000 types of properties [29].

However, the research conducted by the Materials Project in the United States and the Novel Materials Discovery Laboratory in Europe has led to a remarkable shift in how new polymers are discovered. Therefore, some aberrations of present databases still obstruct the extensive use of polymer informatics, this is because numerous data resources are not entirely within reach and are small as compared to the general fields of materials or AI research.

Another reason is the inadequacy of databases comprising processing details or experimental information. Since the development of Polymer Informatics is still in its early state, more research work is in progress for the continuous accumulation of data [27].

8 Machine Learning in Polymer Informatics

To facilitate the discovery, design, and optimization of polymers by harnessing data-driven techniques, machine-learning serves as a tool. Machine learning is an important sub-field of Artificial Intelligence (AI) that enables the computer to learn from data and improve their prediction without being clearly programmed [7, 27].

Earlier, computer was responsible for implementing the hard-coded algorithm designed by field experts which only partially captures the true nature of intelligence. In Polymer chemistry, machine-learning is exponentially being utilized to link chemical structures to macroscopic properties of polymer structures that facilitate to enhance the specific properties. To promote this, a chemical dataset must be translated into machine-readable descriptors. However, machine-learning is acknowledged for its competence to process and examine large volume of data precisely and effectively [9].

Conventionally, machine learning data incorporates collection of input and corresponding output values. However, the learning process of a model encompasses modifying its internal parameters in a function so that it can generate appropriate output values.

9 Machine Learning Algorithms

Algorithm refers to a set of rules that a computer follows to learn patterns from data and make decisions without being programmed for particular tasks. Machine learning approaches depend on different set of credible data sources like experimental, theoretical, computational, and simulated that comprises of adequate information to analyze various targets. However, these targets are analyzed by models which are developed through particular strategies and algorithms.

These models could be in a form such as regression, classification, clustering, dimensionality reduction, or auto-generation form pursuing for the goals like interpretation, prediction, and inference. The overall planning involves preparing and manipulating data and features, building models, evaluating their performance, and incorporating them to relevant problems.

The selection of algorithm depends on the presence of labels in the data and the nature of the task. ML tasks can fall under categories such as supervised, unsupervised, and reinforcement learning; as well as inductive, deductive, and transductive learning. Other prototype include ensemble learning, transfer learning, active learning, federated learning, and deep learning based on neural networks [36].

Machine learning can be divided into three main categories: Supervised, Unsupervised, and Reinforcement learning.

Supervised Learning-In supervised learning, the model is trained on data marked with the target identifier, enabling it to understand patterns and relationship between input features and the target. Supervised learning is specifically useful for tasks including classification and regression [29].

Unsupervised Learning-In unsupervised learning, the model is trained on data which is not marked, enabling it to independently discover patterns or structures in the data. Therefore, this makes unsupervised learning notably good for tasks like clustering and reducing dimensionality [10].

Reinforcement Learning-In reinforcement learning, the model learns through interacting with an environment by performing actions and receiving rewards in the form of a score. This technique is well-suited for learning tasks, robotics, and lastly making decision in real time [3] (Table 2).

Deep Learning:

For complicate polymer datasets, classical ML algorithms like Random Forests, Support Vector Machine (SVM), Principal Component Analysis, and K-means Clustering, deep learning has gained prominence in recent years. For small data collection, defined tabular data, and comprehensible models where transparency is crucial, these methods are highly efficient. From a foundational perspective, deep learning is predicated on Neural Networks (NN). Neural Networks is a mathematical model influenced by the human brain. Just like biological neuron, artificial neural networks incorporates layers of connected neurons. These artificial neural networks however, execute and generalize from data.

Primarily, the input layer accepts unprocessed data, which is then processed via hidden layers where patterns are diagnosed before reaching the final destination that is output layer which formulates predictive results [7].

Steps Involved in Machine Learning Approaches:

Divide the dataset into training, testing and validation to investigate the model's predictive performance. Several techniques can be applied for this task which are hold-Out Validation, K-Fold Cross Validation, Stratified K-Fold Cross Validation, Leave-One-Out or Leave-P-Out Cross Validation, Monte Carlo Cross Validation and Time Series Cross Validation.

Calculate a broad range of molecular descriptors and isolate from formulas, processing and measuring variables. The techniques used in order to build meaningful informational features are Additive Modeling, Encapsulation, Encoding Methods, and Symbolic Assembly.

Develop sturdy models using either a exclusive machine learning algorithm or a pipeline incorporating multiple algorithms. This method can involve repetitive machine learning techniques within a framework, often encompassing evolutionary

Table 2 Categories of machine learning and their characteristics

Category of machine learning	Description	Types/examples
Supervised learning	The model is trained on data marked with the target identifier	– Linear regression – Logistic regression – Support vector machines (SVM)
Unsupervised learning	The model is trained on data which is not labeled	– K-means clustering – Autoencoders
Reinforcement learning	The model learns through interacting with an environment by performing actions and receiving rewards in the form of a score	– Q-learning

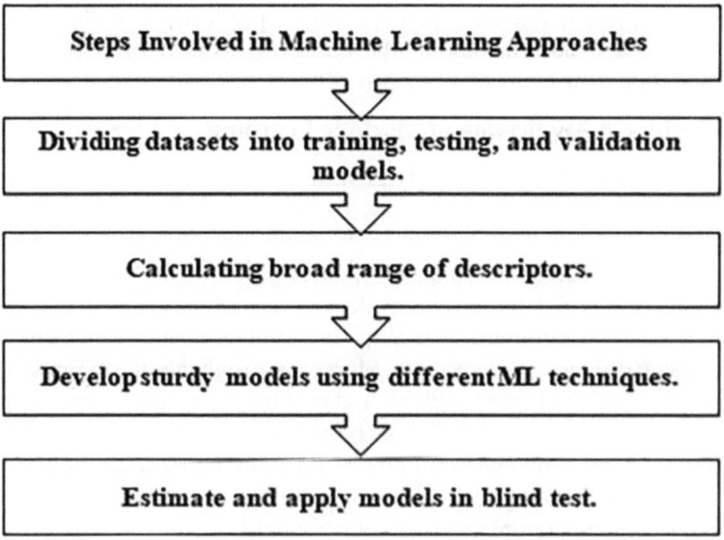

Fig. 1 Flowchart depicting the steps involved in machine learning approaches

methods like Gaussian Processes (GP), Bayesian Optimization (BO) to auto-optimize hyper parameters.

Estimate and employ models in blind tests, virtual screening, optimization tasks and also in rational and inverse design processes [36] (Fig. 1).

9.1 Application of Machine Learning Approaches in Polymer Science

In the field of material science, ML approaches have been widely implemented in synthetic data generation, ingredient identification, three-dimensional structural reconstruction and physical properties determination, in the manufacturing and evaluation of images collected from microscopy, STM, SEM, TEM, FM, and Confocal Microscopy [24].

A core application of machine learning in polymer science includes examining molecular structures via cheminformatics tools like RDKit which aids in translating chemical structures into machine-readable formats such as SMILES strings or molecular fingerprints, which can further be leveraged as inputs for machine learning models. BigSMILES broadens the scope of this feature by allowing the encoding of structural randomness in polymer chains and making it applicable for stochastic polymers [40].

In addition to this, Polymer Genome offers pre-learned models that predict polymer properties thus facilitating the quick assessment of potential polymer material based on their molecular features [7].

Some Examples of Application of Machine Learning

Design of Polymer Dielectrics for High-Energy–Density Capacitors: In energy storage devices, polymer-based dielectric capacitors are frequently applied [5, 13, 28, 31]. The escalating demands for high-energy density capacitors have led to the recent developments in polymer informatics which drives the discovery of new dielectric materials [19]. For high energy density capacitors, generally, good polymer dielectrics has certain requirements to meet for example, enhanced dielectric constant and robust electrical insulation strength. Moreover, to enhance the thermal stability at extreme temperature, polymers with high glass transition temperature are preferred for high-temperature capacitors [35]. The parameters of high glass transition temperature, wide band gap, and enhanced charge injection barriers can be leveraged by combining machine learning approaches to identify polymer candidates customized for extreme high-temperature and high- electric field.

Design of Polymeric Membranes for Gas Separation:

Polymers have proven to be promising candidates for gas separation because of their high surface area [2, 18]. Among polymers, a specific group acknowledged as Polymers of Intrinsic Microporosity (PIMs) has emerged as a subject of considerable interest since the early 1990s. The existing polymer membranes face challenges like low selectivity and physical aging, accentuating the need for the advancement of new types of polymeric membranes. However, finding a promising polymer membrane along with a combination of high permeability and selectivity advancing beyond the trade-off limit represented by Robeson plots for various gas pairs for example O_2, N_2 [25].

Key Stages in Applying Machine Learning to Polymer Science:

The application of machine learning techniques in polymer science typically involves four main stages: Collection of data, Feature and data engineering, Development and examination of models, and Application of models.

Real World Tools:

To foster the incorporation of Machine Learning in Polymer Science, substantial AI tools and platforms are accessible to assist data management, evaluation, and modeling. Open-sources tools are an important part in this ecosystem because they play a key role in fostering transparency, replicability, and approachability encouraging a expanded scientific community to contribute in AI-driven discovery of new materials [7]. Some ML tools and platforms are as follows:

No-Code or Low-code ML Platforms: Teachable machine (FT), Weka (OS), KNIME (OS), Google AutoML (P), Azure ML (P). These tools are ideal for classification, and baseline workflows.

Programming and Implementation Environments: Google Colab (FT), JUpyter Notebooks (OS), Anaconda (FT), Python (OS). These tools are used for interactive coding, script processing, environment organization for data science workflows.

Data cleaning and Formatting: Open source tools like Numpy (OS) and Pandas (OS) simplify data manipulation and numerical operations. These tools and platforms are used for effective handling of arrays, tables, and structured experimental data.

Machine learning Frameworks: Scikit-learn (OS), TensorFlow (OS), these libraries can be used for classical machine learning and deep learning procedures.

Online Resources for Polymer Data: PoLyInfo, CROW, Polymer Genome (FT). These are the data stores of computational and experimental polymer properties.

Integrated Collaboration and Distribution Platforms: GitHub (OS), Zenodo (OS), Hugging Face (OS), and Figshare (OS) [7].

9.2 Challenges and Limitations

The innovation of polymer informatics has been profoundly driven by the consolidation of machine learning techniques, allowing the fast prediction of polymer properties and accelerating the breakthrough of high-performance polymer materials. Nevertheless, the domain falls short of a systematized work process that embraces the accuracy of prediction, uncertainty quantification, and polymer constructability [37]. Along with the triumph of Machine Learning in material science, successful implementation for polymers has been noticed to some extent but wide-scale acceptance has been significantly hindered due to some reasons such as complication in articulating information about complex polymer structure in machine-readable format, inadequacy of available data [29].

Since the incorporation of machine learning strategies into polymer science is highly dependent on the accessibility of organized, high-quality data, various tools and platforms. Table 3 illustrates challenges in polymer science and their respective solutions and methods used to resolve the issue. Although, the escalating accessibility of these datasets, notable obstacles remains unresolved in data-standardization and availability. A variety of studies are compromised by uncoordinated, unreliable data collections. Most frequently, scarcity of enough metadata or eliminating important details about synthesis parameters, characterization techniques, and lastly, measured outcomes. As of now, a number of pioneers in the industry have incorporated AI-based frameworks into their research. For instance, for materials discovery BASF has funded in Artificial Intelligence (AI). In addition to this, one of the world's largest chemical manufacturing companies acknowledged as "Dow Chemical" is investigating machine learning for workflow enhancement [7].

Table 3 Overview of key challenges in polymer science, respective solutions, and the methods used to mitigate them

S. no	Challenges	Solutions	Method used
1	Complex structure of polymer	By converting the information into machine-readable fromat	Fingerprinting
2	Data scarcity	By developing data-respositories and open access tools/platforms	Databases like PoLyInfo, CRIPT (Community resource for innovation in polymer technology)
3	Design space	By automated high-throughput techniques	Robotics and high-throughput experimentation (THE)

9.3 Future Perspectives

The alignment of Artificial Intelligence and Polymer Science is enabling new opportunities, stemming from predictive modeling to self-operating research workflows. Artificial Intelligence will essentially reconfigure how the research based on polymers is conducted, altering from measurable iteration to data-driven predictive modeling [7]. Numerous promising strategies are being developed in response to the challenges and these particular research areas are likely to be the foundation of emerging tools and techniques for the implementation of machine learning as a uniform attribute in the field of polymer science [29].

9.4 Conclusion

The incorporation of data-driven technologies into polymer science represents a paradigm shift in how materials are established, engineered, and customized. This chapter has analyzed the diverse field of polymer informatics, underscoring how the confluence of materials science, machine learning, high-throughput experimentation, and computational simulation is boosting rapid progress in the polymer field. Throughout this chapter, we have studied about the data-driven approaches in polymer informatics. From the development of polymer datasets to the incorporation of different machine learning models for the polymer property prediction and design. Furthermore, we have emphasized on the tools, open access softwares and various platforms which are beneficial to the researchers in order to discover the novel polymer. Important topics like polymer representation, polymer design, and machine learning models have been discussed in order to combat the problems faced by the researchers, varying from predicting glass transition temperatures to designing polymers with targeted thermal, electronic or chemical properties. Implementing data-driven approaches aids in navigating the vast, complicated, multi-dimensional chemical space of polymers. Since the conventional methods are time consuming and highly expensive, thus the advent of new field called Polymer Informatics offers a means to narrow down experimental trials. However, there are certain challenges like data scarcity, irregular data formats. Addressing these challenges will require initiatives in sharing of the data, model interpretability. However, the endorsement of Artificial Intelligence (AI) in polymer science illustrates a remarkable shift for researchers accustomed to conventional methods. For a short period of time, scientists and AI professionals can collaborate and help address the gap. Therefore, in the long run, amalgamating machine learning and artificial intelligence education into polymer science educational programs. In conclusion, data-driven approaches can stimulate polymer design and helps in developing structure–property correlations. In addition to this, these approaches are a paradigm shift in material research.

References

1. An, Y., Singh, S., Bejagam, K.K., Deshmukh, S.A.: Development of an accurate coarse-grained model of poly (acrylic acid) in explicit solvents. Macromolecules **52**(13), 4875–4887 (2019)
2. Barnett, J.W., Bilchak, C.R., Wang, Y., Benicewicz, B.C., Murdock, L.A., Bereau, T., Kumar, S.K.: Designing exceptional gas-separation polymer membranes using machine learning. Sci. Adv. **6**(20), eaaz4301 (2020)
3. Botvinick, M., Ritter, S., Wang, J.X., Kurth-Nelson, Z., Blundell, C., Hassabis, D.: Reinforcement learning, fast and slow. Trends Cogn. Sci. **23**(5), 408–422 (2019)
4. Chandrasekaran, A., Kim, C., Venkatram, S., Ramprasad, R.: A deep learning solvent-selection paradigm powered by a massive solvent/nonsolvent database for polymers. Macromolecules **53**(12), 4764–4769 (2020)
5. Chen, L., Pilania, G., Batra, R., Huan, T.D., Kim, C., Kuenneth, C., Ramprasad, R.: Polymer informatics: current status and critical next steps. Mater. Sci. Eng. R. Rep. **144**, 100595 (2021)
6. Chew, A.K., Afzal, M.A.F., Chandrasekaran, A., Kamps, J.H., Ramakrishnan, V.: Designing the next generation of polymers with machine learning and physics-based models. Mach. Learn. Sci. Technol. **5**(4), 045031 (2024)
7. Ferji, K.: Basic concepts and tools of artificial intelligence in polymer science. Polym. Chem. **16**(21), 2457–2470 (2025)
8. Gakh, A.A., Burnett, M.N.: Modular chemical descriptor language (MCDL): composition, connectivity, and supplementary modules. J. Chem. Inf. Comput. Sci. **41**(6), 1494–1499 (2001)
9. Ge, W., De Silva, R., Fan, Y., Sisson, S.A., Stenzel, M.H.: Machine Learning in polymer research. Adv. Mater. 2413695 (2025)
10. Glielmo, A., Husic, B.E., Rodriguez, A., Clementi, C., Noé, F., Laio, A.: Unsupervised learning methods for molecular simulation data. Chem. Rev. **121**(16), 9722–9758 (2021)
11. Gurnani, R., Kuenneth, C., Toland, A., Ramprasad, R.: Polymer informatics at scale with multitask graph neural networks. Chem. Mater. **35**(4), 1560–1567 (2023)
12. Huan, T.D., Ramprasad, R.: Polymer structure prediction from first principles. J. Phys. Chem. Lett. **11**(15), 5823–5829 (2020)
13. Huan, T.D., Boggs, S., Teyssedre, G., Laurent, C., Cakmak, M., Kumar, S., Ramprasad, R.: Advanced polymeric dielectrics for high energy density applications. Prog. Mater Sci. **83**, 236–269 (2016)
14. Huan, T.D., Mannodi-Kanakkithodi, A., Kim, C., Sharma, V., Pilania, G., Ramprasad, R.: A polymer dataset for accelerated property prediction and design. Sci. Data **3**(1), 1–10 (2016)
15. Kuenneth, C., Rajan, A.C., Tran, H., Chen, L., Kim, C., Ramprasad, R.: Polymer informatics with multi-task learning. Patterns **2**(4) (2021)
16. Lee, J., Seko, A., Shitara, K., Nakayama, K., Tanaka, I.: Prediction model of band gap for inorganic compounds by combination of density functional theory calculations and machine learning techniques. Phys. Rev. B **93**(11), 115104 (2016)

17. Lin, T.S., Coley, C.W., Mochigase, H., Beech, H.K., Wang, W., Wang, Z., Olsen, B.D.: BigSMILES: a structurally-based line notation for describing macromolecules. ACS Cent. Sci. **5**(9), 1523–1531 (2019)
18. Low, Z.X., Budd, P.M., McKeown, N.B., Patterson, D.A.: Gas permeation properties, physical aging, and its mitigation in high free volume glassy polymers. Chem. Rev. **118**(12), 5871–5911 (2018)
19. Mannodi-Kanakkithodi, A., Chandrasekaran, A., Kim, C., Huan, T.D., Pilania, G., Botu, V., Ramprasad, R.: Scoping the polymer genome: a roadmap for rational polymer dielectrics design and beyond. Mater. Today **21**(7), 785–796 (2018)
20. Mogurampelly, S., Borodin, O., Ganesan, V.: Computer simulations of ion transport in polymer electrolyte membranes. Annu. Rev. Chem. Biomol. Eng. **7**(1), 349–371 (2016)
21. Otsuka, S., Kuwajima, I., Hosoya, J., Xu, Y., Yamazaki, M.: PoLyInfo: polymer database for polymeric materials design. In: 2011 International Conference on Emerging Intelligent Data and Web Technologies, pp. 22–29. IEEE. (2011)
22. Patra, T.K.: Data-driven methods for accelerating polymer design. ACS Polym. Au **2**(1), 8–26 (2021)
23. Popescu, M.C., Balas, V.E., Perescu-Popescu, L., Mastorakis, N.: Multilayer perceptron and neural networks. WSEAS Trans. Circ. Syst. **8**(7), 579–588 (2009)
24. Ragone, M., Shahabazian-Yassar, R., Mashayek, F., Yurkiv, V.: Deep learning modeling in microscopy imaging: a review of materials science applications. Prog. Mater Sci. **138**, 101165 (2023)
25. Robeson, L.M.: The upper bound revisited. J. Membr. Sci. **320**(1–2), 390–400 (2008)
26. Rogers, D., Hahn, M.: Extended-connectivity fingerprints. J. Chem. Inf. Model. **50**(5), 742–754 (2010)
27. Sha, W., Li, Y., Tang, S., Tian, J., Zhao, Y., Guo, Y., Cheng, S.: Machine learning in polymer informatics. InfoMat **3**(4), 353–361 (2021)
28. Sharma, V., Wang, C., Lorenzini, R.G., Ma, R., Zhu, Q., Sinkovits, D.W., Ramprasad, R.: Rational design of all organic polymer dielectrics. Nat. Commun. **5**(1), 4845 (2014)
29. Struble, D.C., Lamb, B.G., Ma, B.: A prospective on machine learning challenges, progress, and potential in polymer science. MRS Commun. **14**(5), 752–770 (2024)
30. Stubbs, C.D., Kim, Y., Quinn, E.C., Pérez-Soto, R., Chen, E.Y.X., Kim, S.: Predicting homopolymer and copolymer solubility through machine learning. Dig. Discov. (2025)
31. Tan, Q., Irwin, P., Cao, Y.: Advanced dielectrics for capacitors. IEEJ Trans. Fundam. Mater. **126**(11), 1153–1159 (2006)
32. Tao, L., Varshney, V., Li, Y.: Benchmarking machine learning models for polymer informatics: an example of glass transition temperature. J. Chem. Inf. Model. **61**(11), 5395–5413 (2021)
33. van den Hurk, R.S., Pirok, B.W., Bos, T.S.: The role of artificial intelligence and machine learning in polymer characterization: emerging trends and perspectives. Chromatographia 1–7 (2025)
34. Venkatraman, V., Alsberg, B.K.: Designing high-refractive index polymers using materials informatics. Polymers **10**(1), 103 (2018)
35. Wu, C., Deshmukh, A.A., Li, Z., Chen, L., Alamri, A., Wang, Y., Cao, Y.: Flexible temperature-invariant polymer dielectrics with large bandgap. Adv. Mater. **32**(21), 2000499 (2020)
36. Xie, C., Qiu, H., Liu, L., You, Y., Li, H., Li, Y., An, L.: Machine learning approaches in polymer science: progress and fundamental for a new paradigm. SmartMat **6**(1), e1320 (2025)
37. Xu, J., Liu, G., Guo, R., Jiang, M., Luo, T.: POINT $^{2} $: a polymer informatics training and testing database. arXiv preprint arXiv:2503.23491 (2025)
38. Yan, C., Li, G.: The rise of machine learning in polymer discovery. Adv. Intell. Syst. **5**(4), 2200243 (2023)
39. Yan, C., Feng, X., Li, G.: From drug molecules to thermoset shape memory polymers: a machine learning approach. ACS Appl. Mater. Interf. **13**(50), 60508–60521 (2021a)

40. Yan, C., Feng, X., Wick, C., Peters, A., Li, G.: Machine learning assisted discovery of new thermoset shape memory polymers based on a small training dataset. Polymer **214**, 123351 (2021b)
41. Zhu, S., Lempesis, N., in 't Veld, P.J., Rutledge, G.C.: Molecular simulation of thermoplastic polyurethanes under large compressive deformation. Macromolecules **51**(22), 9306–9316 (2018)

Machine Learning Innovations in Food Packaging: Enhancing Safety and Sustainability

Deepika Goswami, Hema Arya, Anzala Mumtaz, Kiran Negi, and Sandeep Dutt Mandoli

Abstract Machine Learning has made a significant impact on the food packaging industry by predicting the spoilage of food and monitoring the food items. This chapter involves the use of AI and machine learning techniques and its evolution leading to sustainable development of packaging materials resulting in reducing carbon footprint. Smart and intelligent packaging has also made the machine learning incorporation easy and provides better tools and techniques which benefits the food packaging industry. It uses techniques such as biosensors, IoT based sensors, RFID tags, barcodes and bioplastic to make the food free from contaminants which are capable of causing food-borne disease by analysing change in pH, moisture, gases, etc. ML packaging materials are the alternative of plastic based packaging. Leading companies like Coca-Cola uses machine learning in food packaging process and have solved various problems that other companies might be growing through. Future research should focus upon personalized packaging, insuring more safety of consumers privacy.

Keywords Machine learning · IoT based sensors · RFID tags · Sustainable packaging · Bioplastic

1 Introduction

Machine learning is a great innovation in food packaging industry which consists of finding and making models and algorithms by use of which computer gain information by providing input data [37]. Food industries administrating machine learning in packaging has resolved various problems majorly, packed food causing more than 200 diseases such as diarrhea, vomiting and in severe cases cancer as per WHO [56]. The global population approaching 10 billion, the demand for food, food supplies, food packaging and preservation techniques for packaged food for extended periods of time has become one of the main concerning topics [23, 33]. Packaging is the

D. Goswami · H. Arya (✉) · A. Mumtaz · K. Negi · S. D. Mandoli
Department of Zoology, DSB Campus, Kumaun University, Nainital, India
e-mail: aryahema557@gmail.com

B. Sharma et al. (eds.), *AI-Driven Polymer Design*, Engineering Materials,
https://doi.org/10.1007/978-981-95-8353-9_6

systematic process used for the transportation, storage, use and sale of raw. It offers protection from microorganisms, CO_2 (carbon dioxide), O_2 (oxygen), humidity, and other gases [12, 38].

Packaging of food is broadly categorized into two types: active packaging and smart packaging systems. Active packaging gives information regarding packed food and the suitable condition it requires to be protected from spoilage and extend its freshness for a prolonged period of time [51]. For example, the implication of active packaging of strawberries is recorded to increase their freshness and shelf life [12, 53].

Smart packaging systems have two types of devices: data carrier type and indicator type. Data carrier devices include barcodes, radio frequency (RFID) tags, etc. Whereas indicator-type devices are used to analyze any changes in temperature with respect to time and gas indicators [57]. For maintaining the freshness of food, the time–temperature indicator uses fadable inks for printing labels, which tend to fade as time passes. So if any food item is kept for long, this ink will disappear, indicating the food is no longer fresh and therefore must not be consumed. Gas indicator Provide information on whether packed food contains gas, which will harm the food's freshness, and therefore make the consumer aware regarding the safety of the packed food [25]. A few other measures that have been taken by the packaging industry for the safety of their consumers are the sticker indicator, also known as the sticker-based sensor, which is used for broiler chicken testing. If the sticker's color is no longer the same (red) as it was at the time of packing and turns tawny, it informs the consumer that it is no longer safe to consume such type of meat. The sticker-based sensor technique uses the methyl red fixating bacterial-derived cellulose film over the broiler chicken [24].

Packaging of food has been an ancient ritual noted in every culture. Ancient food packaging includes packing food with leaves, animal skin, clothes as well as utensils. After the end of ancient packaging came traditional packaging which include manual packaging. Traditional packaging style was preferred few decades ago which had few benefits but it had major drawbacks, including not able to maintain the freshness of food and was not environmental friendly [2].

The use of plastic for packaging prevailing for more than a few decades, as it offered benefits such as packaging for food items such as soups, curries, etc., which could not be packed in paper bags and cardboard. The wet food items could either be packed in bottles, which was expensive and also caused more harm to the environment. Plastic ruled the packaging industry, but due to its resistance to natural decomposition, it failed in becoming the best packaging material [1]. As a result, biotechnologist introduced biopolymer, a decomposable polymer that was environmental friendly, to the world. A biopolymer is made up of natural materials or bioengineered materials [47]. The use of biopolymer in the packaging industry has reduced the greenhouse gas emissions as it is made up of renewable resources. Biopolymer may take over plastic in food packaging in a few years, it will be possible only if it is cost-efficient, doesn't get damaged after coming in contact with water, which affects its shelf life, and offers better durability than plastic.

These problems gave rise to Artificial Intelligence (AI)-driven machine learning technology, which facilitates the system to learn from data and improve its performance on its own without direct human programming. For example, facial recognition, personalized recommendation systems, etc. It is also important as it easily performs the complex procedures within a short period of time that are time-consuming for humans. Machine learning, therefore, is a boon to our food industry as it continuously improves the functionality and authenticity of processes and could be used to generate new and innovative compounds and facilities [14]. Food and drink packaging industry comprises 55–65% of the $130 in United States [4]. Growing Indian packaging industry is expected to make approximately $143 billion by the year 2029. This growth is due to the growing populations need and their demand for processed food, increase in daily wages and changing lifestyle [5].

The technological advancement in machine learning allowed mechanization of monitoring and advancement of predictive analytics ensuring efficacy of food safety and immediate response to potential risk [35]. Artificial intelligence gave rise to machine learning, which helps to gain insights and optimize the system from prior experience without any human assistance. The main purpose of the advancement in programming in computers is to obtain data and utilize it independently.

AI driven machines can evaluate nutritional requirement and choice of an individual to create products specifically built for consumers need [40]. It provides the information regarding the product quality and all the food ingredient, preservatives, etc., transparent to the consumer or else the food companies may add ingredients that has been banned due to their hazardous effect on human. For example—the Sanlu Milk Melamine powder case from China in 2008 highlighted serious concerns regarding food safety, public health, and the health of newborn babies that we often overlook. This milk resulted in the death of most of the babies. Another such case from China is its dominant meat distribution company, Shaunghui Group, which added clenbuterol hydrochloride to their meats, which is banned as it is potentially harmful for the health of the consumers [45, 51].

2 Overview of Machine Learning in the Food Industry

Basic concepts of machine learning

Machine learning is a technique which consists of finding and making models and algorithms by use of which computer gain information by providing input data. Machine learning can also be defined as measure of prediction and analysis by providing information to the computer [37]. The application of such emerging technology have supported the inspection of food and supply chain management [17].

Historical evolution and current trends

Packaging plays a crucial role in the food packaging industry, which works on protecting food items from contamination and therefore maintains their freshness for a long period of time. The contaminants could be microorganism, chemical exposure, gases, moisture, nonsuitable temperature etc. Different packaging material have gained recognition in food packaging industry such as paper and paperboard which is used 30–35% as they are easily recycled and also environmental friendly, for the long term use and safety of food aluminium and tin are used 10–15%. Biodegradable packaging material such as biopolymer is used upto 5–10% [5].

3 Current Challenges in Food Packaging

Food safety risks (contamination, spoilage)

Globally estimated lost of fresh fruit and vegetable is upto 50% due to spoilage, wastage in the food supply chain [34]. Food safety is the ultimate aim food industries as well as consumers. From the manufacturing to the delivery of food items are prone to come in contact with contaminants and spoilage. There are various types of food items which gets spoiled with different contaminants. For example—Bakery items are prone to get spoil if exposed to microorganism, moisture, chemicals etc. The 1980s marked the era of bakery product being spoiled by bacteria namely, *Bacillus* spp. like *subtilis* and *mesentericus* [32].

Meat and meat products: Meat industry is one of the most powerful industry ruling 21% in major industries. Meat is sold either raw or cooked by packing in a suitable packaging material such as glass, plastic, bio-based materials metal containers etc. The main contaminants of meat are microorganisms, chemical exposure, and rancidity. The modern approach in packaging industry to resolve the problem are using vacuum packaging in which involves packing meat with plastic and removing air from it. Once packed with vacuum packaging it looks like the plastic is too tight for the meat. Another approach is Modified Atmosphere Packaging (MAP) which is popular among consumers as it serves the right colour of the food, extends the freshness period of the food and helps in reducing wastage of food [7].

Fruits and vegetables Major contaminants of fruits and vegetables are Aspergillus, Penicillium, Saccharomyces and Candida, which works by producing hazardous substance, fermenting sugar and odd flavor in fruits and vegetables. These result in many health hazard in humans [21]. For example apple prone spoilage from *A. niger* and *P. expansum* was detected with gas sensor [16].

By evaluation extensive datasets, ML algorithms can determine biological food contamination, beforehand alert food deterioration, and make traceability more reliable of ASF supply chains [35, 41].

4 Machine Learning for Food Safety in Packaging

Predictive models for contamination detection

Contamination refers to the process by which impurities are present in a clean material and becomes unfit for the consumption. For example—bread becomes contaminated when it comes in contact with the fungus, bacterial, moisture, eaten by rats or other animals, etc. Machine learning have successfully generated models that are use to predict food spoilage and monitor the foot items so that they don't come in contact with any contaminant [44]. There are various technologies in the machine learning for monitoring and predicting shelf-life, spoilage and contamination of food which are as follows.

Machine learning algorithm which can monitor food items and evaluate its shelf-life have various models such as-IoT based sensors. Internet based sensors are the sensors which make use of internet foe data collection and distribution therefore continuously creates enormous amount of data sets. These datasets consists of information of the food temperature, pH, moisture content, microorganism growth, chemical exposure. Intellistore is one of the IoT and artificial intelligence based storage device use to track and administer easily spoiled items [3, 36].

Computational intelligence based on deep learning and the YOLO (You Only Look Once) framework systematically enables fast-acting machine learning models that evaluate the slots and sections of images, delivering exceptional results in identification and organization. These models help in acquiring instantaneous data which gives information like packaging irregularities which promotes producing high quality food packaging and enhance precision and durability in real world use [22]. For example-apple sorted and graded using BiseNet V2 use to maintain quickness and correctness (Bilateral Segmentation Network Volume 2) and YOLO V4 (You Only Look Once Volume 4) [27].

YOLOV8 application in biscuit industry by classifying high quality biscuits and maintaining the quality. YOLOV8 models use image recognition technology and remove defected biscuits [22].

ML in microbial growth and spoilage prediction

1. **Deep Learning and Image Recognition**

Image recognition technology is the machine learning innovation that helps by analyzing data based on the tracking difference in the image of food product such as an 1st day picked apple image and the image of the same apple on the 5th day. Image recognition is the best technique for detecting spoilage of food products with the help of continuous filming [3, 27].

Real-time monitoring using sensors and ML algorithms

2. **Sensors**

Sensors are the intelligent monitoring instruments which keeps monitoring the quality of food and ensure its safety. They offer safety of food the by detecting change in

Sensor Type	Sub-Types / Examples
Gas Sensor	- MOX (Metal Oxide) gas sensor- SAW (Surface Acoustic Wave) gas sensor- Electromagnetic gas sensor- Colorimetric gas sensor
Temperature Sensor	- Thermochromic ink
Identification / Tracking Sensor	- RFID tags

Fig. 1 Two major types of sensors

physical, chemical, environmental parameter of the standard condition of the food item [31].

3. **Gas Sensor**

Gas sensors are the analyzing tool used for the detection of the gases which depict spoilage of food items. Gases that depicts spoilage are CO_2, O_2 ethylene etc. A preferable sensor is the one which can easily be used and can detect spoilage in a very short time and offer low retention. The minimum amount of gas a sensor can detect is known as the limit of detection (LOD) [29].

Sensors are of different types on the basic of function and quality such as MOX (Metal Oxide Sensor) gas sensor, colorimetric gas sensor, electrochemical gas sensor and SAW (Surface Acoustic Wave) gas sensor. MOX gas sensor are used in tracking parameter of water, food, disease and mechanical operation. SAW gas sensor works on the principle of preferred membrane coated on the surface of sensor absorbs particular gas [30] (Fig. 1).

4. **Temperature and Humidity Sensors**

When a food container has a temperature sensor it helps in monitoring the adequate temperature and educating the consumer regarding the product quality. Thermochromic ink dot is the widely used time–temperature indicator. For example-UK and USA use plastic container used for the making pancakes consists of thermochromic ink dot that changes color revealing the syrup is ready to be consumed after heating in microwave [25].

5. **Machine Learning for Sustainable Packaging**

Sustainable packaging refers to measures taken to make food packaging material in such way it reduces the environment impact and can be easily recycled. For example-bioplastic made from sea weed, starch poly lactic acid, cellulose, protein etc. [18, 28].

Design of biodegradable and recyclable packaging

Bioplastic

Bioplastic is a new biodegradable plastic that is made up of plant-based renewable resources. It has been brought to use to reduce the use of harmful plastic and reduce the environmental pollution through it [42]. Bioplastic has been invented in twentieth century which aims to minimize agricultural waste and convert agricultural waste to bioplastic. It is majorly made up of carbohydrate combined with other material. The manufacturing of bioplastic depends upon the agricultural example-potato peel, seaweed *Kappaphycus alvarezii* [15, 39].

Bioplastic is categorized on the basis of biodegradability and material used such as—bio-based plastic and biodegradable plastic. Bio-based plastic doesn't refer to plastic that is biodegradable but it means bioplastic that is made up of biological material for preparing bioplastic for example—sugarcane palm seed. Biodegradable plastic does not necessarily mean has to be made from plant but it means bioplastic that can be biodegradable and can be made from starch and Polylactic acid (PLA) [1, 28].

Bioplastic has few demerits such as it can easily absorb water as a result can be destroyed and can damage the food product. It is also weak in strength therefore no heavy food items can be packed by using bioplastic [48].

Machine learning is used in making these bioplastic sheets for packaging. Figure 2 shows hows the bioplastic is made by using machine learning.

Smart labeling and traceability

Internet of things (IoT) based sensors by using Fourier transform infrared spectroscopy (FTIR) can easily detect the meal quality and can monitor the spoilage [19]. It can also categorize meat into halal and non-halal meat. Halal meat is the meat

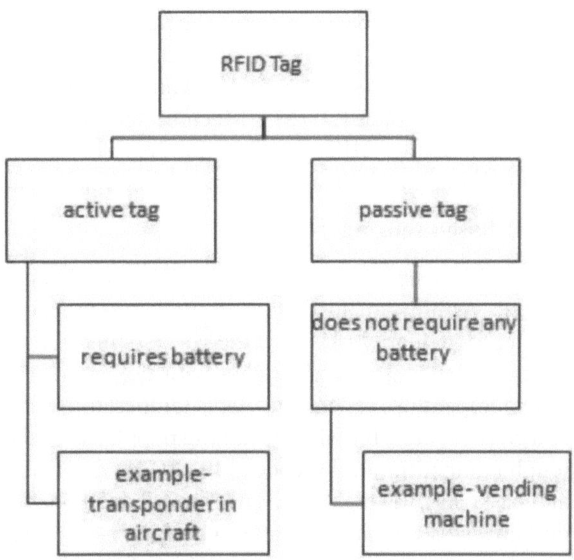

Fig. 2 Shows types of RFID (radio frequency identification) and its example

which can be consumed under Islamic dietary laws whereas non-halal meat includes pig meat and its by products are non-consumable. FTIR is a new advanced technique generate from IR (infrared radiation) spectroscopy which works in 400–4000 cm^{-1}. It is one of the powerful technique used for the differentiation and examination of food especially meat such as lard, chicken and cow [52].

6. **Smart Packaging Systems**

The increasing food demand, has been a significant reason for the rise in food-borne diseases, which pose a threat to public health. The USDA reports that the cost of dealing with food-borne illnesses grew significantly over five years, escalating from \$15.5 billion in 2013 to \$17.6 billion in 2018 [55]. Indicating the rising economic burden of demanding safe processing and packaging of food, ensuring better health of the people consuming processed food [43].

Technology in the food industry has made it easier to detect and report issues related to the packaging of food items. A biosensor is a data-driven device consisting of a transducer and bioreceptor, which convert the resulting interaction into a measurable, continuous output signal. Biosensors are of different types, such as enzymatic, microbial, chemical, optical, and electrochemical, and IoT-based (internet of things) biosensors [17]. One of the prominent strategies is inculcating an advanced food technology system that takes advantage of IoT-based biosensors, facilitating the assessment of safety risks throughout the entire food supply chain, starting from storage to logistical handling. When incorporated directly into food, they help in providing continuous real-time conditions of the food that is packaged in order to signal the presence or absence of food pathogens and pollutants [10]. Implementation of IoT biosensors in the packaging industry, especially in food packaging, recognizes and evaluates the nutritional integrity and degradation of food purity and helps in handling food freshness and maintaining its condition to avoid spoilage. Moreover, IoT-based biosensors continuously examine the parameters such as temperature, humidity, and pH of food during shipment, and if any changes in these parameters are recorded, these IoT-based biosensors send an immediate signal indicating the risk to the safety of the food product [20, 51]. Machine learning algorithms associated with hyperspectral imaging (HIS) deliver a progressive approach for organizing plastic into food-contaminated plastic and non-contaminated plastic [46] (Table 1).

RFID-based sensing

Radio Frequency Identification (RFID) based sensing used for the classification, logistic visibility of food items to make them safe for the consumption and maintaining food quality [31]. It is the wireless inspection system that monitor particular items and store and update data with the use of radio wave [50, 57]. RFID is an time–temperature sensor (it detects temperature variation along time) which is used as a chipless tags. It is slowly gaining popularity in the packaging industry and giving tough competition to barcodes [49]. RFID can be broadly divided into type: active and passive RFID tags. Active tags work due to battery therefore they are dependent on battery as an energy source whereas passive tags does not require any battery. Passive tags operation under ultra-high frequency (UHF) used for the production,

Table 1 New technology in food packaging

New technology in food packaging	Definition	Function	Advantages	Disadvantages
Internet of things (IoT)-based sensors	Use of internet-enabled sensors for tracking and prediction of changes in food quality	Monitoring of pH, temperature, moisture, and microbial growth	Maintains food freshness and alerts consumers to quality changes	Data privacy concerns
Nanotechnology	Use of nanoparticles in packaging materials	Enables active, smart, or intelligent packaging	Extends shelf life and reduces food wastage	Toxicity and long-term safety still uncertain
Biopolymers	Use of natural materials to produce biodegradable polymer-based packaging	Reduces plastic usage through biodegradable packaging materials	Renewable, non-toxic, environmentally safe	Relatively expensive

monitoring in food supply chains have unique electronic product code which makes them different from each other. RFID tags benefits the food industry by being cost effective and monitoring products but it cannot be used by every supply chain as their are still hurdles in implementing and expanding these sensor-driven system [3, 6, 50]. Although biosensors make profit for the food packaging industry but it could create some dangerous problems as well. Biosensors which fetch the data for the safety of food could be hacked and manipulated. IoT based biosensors are expensive therefore most of the food blocks cannot afford it.

E-Nose detection-used to detect the gases and volatile compounds which could easily tell spoilage of fruit, vegetable and other food items. One of the application of this technique was for the detection of kiwi fruit spoilage [9].

5 Machine Learning Applications

In recent years, Machine learning (ML) has become a driving force in food packaging. It facilitates data-driven formulations to persistent limitations in safety, quality assurance, and sustainability. Machine learning approaches can be employed for harvesting information, make predictive result, and detect patterns. For example, predicting the shelf life of temperature-sensitive products, a multiple non-uniform regression model has been developed, derived from thermal sensor data to identify a correlation between temperature and loss of shelf life [8].

Computational learning systems, or machine learning systems, can monitor underlying structures in the data and prognostic datasets to aid in determining outcomes and inferences. The food industry widely uses fruits for various purposes, such as juice, dried form, candies, etc. To retain their qualities for a longer period of time, they often preserve fruits. The introduction of machine learning, preservation of fruits has become much easier and more time-efficient. It facilitates the collection and processing of data from various fruits to ascertain the preferable condition for the fruits to maintain their freshness and prevent spoilage. Machine learning (ML) uses data such as chemical properties (pH level, glucose and fructose level, organic acids, ascorbic acid, etc.) and environmental conditions (humidity, exposure to light radiation, temperature, etc.) that shape the rate of microbial decay and shelf life of fruit. The ML algorithm makes anticipatory models that are able to estimate ideal conditions for the maintenance of fruit quality. It can also estimate the dimension, pigmentation, morphology, surface texture, and even ripening period of the fruit. By estimating this data can reduce waste generation, improve expiry duration, and determine fruit fit for consumption and therefore give valuable information regarding the quality of fruit to its consumers [43].

6 Case Studies

A study was conducted which employed combination of Machine learning methods like **CNN-SVM** (Convolutional Neural Network and Support Vector Machine). The **CNN-SVM** combination is frequently implemented to leverage each method. CNN works as an effective tool for subset selection. On the contrary, SVM acts as a powerful classifier. This technique was employed to assess the freshness of Salmon over the period of IoT-empowered cold storage and categorize every salmon sample based on the freshness levels [11].

Another study was conducted by Aytaç and Korçak, where they inspected the reliability of Naïve Bayes based on Bayes theorem purposed by Thomas Bayes and decision trees for the prediction of restaurant demand. The models underwent training on waste bin weight data, progressive sales data, and exogenous event data extracted from social media platforms and internet which could influence demand. The learning data were labelled physically with a service-level indicator. After the learning was done, the model was able to predict the production service level without any human assistance, implying the arriving customer did not have to wait for food to be prepared, whilst reducing the amount of food waste produced due to the short lifetime of the product. Furthermore, the study employed an unsupervised learning method to execute outlier recognition [8].

Machine learning has addressed many challenges in the food industry, ranging from minimizing the need for human assistance to improving packaging for food products [26]. It has brought major changes in designing the new product, also known as the **Product Development Process (PDP)**, divided into three phases: conceptual, embodiment, and detail. Modular deviation plays a vital role in the conceptual phase,

Table 2 Shows different food items, their contaminant and machine solution

Food	Contaminant	Machine learning solution
Bread	Moisture, microorganism (*Bacillus subtilis* and *Bacillus mesentericus*)	Biosensors
Meat	Microorganism (*E. coli*, *Salmonella*), temperature	RFID tags
Milk	Plastic, microorganism (*Mycobacterium bovis*)	RFID tags, time–temperature biosensor
Fruits and vegetables	Gases, temperature, microorganism	Image recognition technology

where the highest degree of freedom is present along with interfaces between devices [13].

The canning industry which used liquid known as sauce in the packaging of fish with the help of vacuum sauce filling machine which was designed to reduce the amount of sauce lost in the process. There has been other advancement as well such as using automated handling system for packaging tasks. Different innovation took place trying to give the best module for packaging of fish. The machine learning adapted the morphological matrix which contributed the most in the fish packaging [13] (Table 2).

7 Case Studies and Industry Application

Big brand companies have started using machine learning to make profit and without wastage of the food products such as:

Coca cola started using IoT sensor for monitoring and tracking the condition such as temperature and moisture content of the cold drink throughout the manufacturing journey. It also helps the product to reach its consumers in the best condition without changing its taste. The application of IoT based sensors generate large amount of real-time data which allow modification in supply chain process. Reduction in the rate of product wastage and sustainable working enhancing the productivity [20].

8 Ethical Considerations

8.1 Data Privacy

Data security and privacy is one of the main concern while using IoT based sensors used for food packaging as it is prone to hacking, data manipulation. These safety concerns are due to the biosensors used to collect data, data transfer and other abiotic factors influencing microbial growth. AI when used with biased data may result in serious concern in food safety. The use of machine learning in food packaging industry resulted in loss of manual force and decrease in employment rate [54].

9 Challenges in Implementing Machine Learning in Food Packaging

Machine learning in food packaging usage in food industry has grown significantly over the years but several critical challenges limit its full-scale implementation. These challenges are technical, legal, economic, ethical, implementable and social.

Cost and scalability of intelligent packaging technologies is the most important challenge which must be addressed. Integrating machine learning with IoT sensors, RFID tags, and biosensors increases the cost of the packaging for the industry as the machine are expensive. Although bioplastic reduce the carbon footprint and help in sustainable food packaging but due to its high cost of manufacturing and less popular to the consumers it becomes a serious challenge for the packaging industry [42].

Data privacy and cybersecurity. Smart sensors and IoT-based biosensors gathers personal information which can easily be accessed by the hackers. These issues raises concern in logistics, and consumer behavior. Without encryption and cybersecurity protocols, they make the system sensitive to hacking and use of this information for wrong means which poses a threat to consumer safety but also raises concerns about the misuse of proprietary industrial data [17].

Lack of standardization and regulatory frameworks. The growing sector of AI and ML development in food packaging has overtaken current food safety and packaging gap in most of the countries. The policy gap causes confusion around the legal validity of ML-based packaging systems, especially shelf-life prediction, biodegradable sensors, and personalized labeling [45].

Technical complexity and integration issues persists to remain a challenge. Many food manufacturers although use machine learning for packaging but doesn't truly know how to make use of it to the fullest so hiring a well trained staff needs more money and therefore will increase economic burden [35].

10 Future Prospects and Research Directions

The innovation of machine learning technology in food industry has made significant changes starting from replacing plastic to provide safety of food from contaminants. Growing use of AI has driven industries better accuracy and affordability and helped in adopting profitability and efficiency. Smart packaging has resolved issues related to maintaining freshness of fruits, vegetables and other processed food items. The biodegradable polymer-bioplastic has reduced carbon footprint but due to its high cost it cannot be used by many industries. IoT based sensors are widely used a due to easy handling therefore consumers can easily access the information of the food they are consuming. Future research should work on making cost-efficient packaging material. It should aim in making IoT sensors provide more security to the consumer privacy. The knowledge gap of employees working in industries must be addressed consisting of machine learning for packaging as a result it could be easily used and help in entrepreneurship.

11 Conclusion

This chapter describes how machine learning is transforming the food packaging industry by making the major changes required such as reducing the use of plastic, monitoring the food item and predicting its expiry date, giving honest information about the product and introducing environmental friendly packaging products. Machine learning algorithm supports predictive models that detect contamination caused by microorganism to sensor development for monitoring change in pH, moisture, gas, temperature therefore machine learning has become an important part for improving packaging dependability and improving the health of consumers by reducing the risk of food-borne illness. Machine learning has not only increased the profit of the food industries but has also reduced the agricultural waste transforming it to produce sustainable packaging through the development of packaging material such as bioplastic and biopolymer which are alternative to conventional plastic packaging. Transformation of machine learning can design packaging models by fulfilling the design requirements to be modular, adaptive, and responsive to time-sensitive data. Various techniques including modular matrix helps in the food industries like tuna canning, biscuit and potato chips. Sensors development such as IoT based, time–temperature sensor and gas sensor have made the prediction of food spoilage fast and accurate. Deep learning, 3D packaging, face recognition, barcodes, RFID tags have made the consumer access to the information regarding the food item they are consuming by just using their phones. Machine learning innovation although has many benefits but it demerits are severe such as leaking of data, hacking and fraud. Machine learning in food packaging is increasing consistently and it paves way awareness of cybersecurity. In conclusion to this chapter machine learning aims to bring big changes in future and offer more safer and quality food products.

References

1. Abhitha, S., Amulya, P., Dubey, A., Praveen, B.V.S., Madhuri, P.: Comparative analysis of bioplastic sheets derived from biowaste using various plasticizers. Arab. J. Sci. Eng. 1–13 (2025)

2. Awlqadr, F.H., Altemimi, A.B., Omar, A.M.A., Saeed, M.N., Qadir, S.A., Faraj, A.M. et al.: Advancing sustainability in fruit and vegetable packaging: the role of nanotechnology in food preservation. *eFood* **6**(3), e70060 (2025)
3. Bongarde, D., Pandit, S., Pandit, H.: Use of Machine Learning and Artificial Intelligence in Food Spoilage Detection (2024)
4. Brody, A.L., Bugusu, B., Han, J.H., Sand, C.K., McHugh, T.H.: Innovative food packaging solutions. J. Food Sci. **73**(8), 107–116 (2008)
5. Cai, J., Zhang, D., Zhou, R., Zhu, R., Fei, P., Zhu, Z.Z., et al.: Hydrophobic interface starch nanofibrous film for food packaging: from bioinspired design to self-cleaning action. J. Agric. Food Chem. **69**(17), 5067–5075 (2021)
6. Claucherty, E., Cummins, D., Aliakbarian, B.: RFID unpacked: a case study in employing RFID tags from item to pallet level. Electronics **14**(2), 278 (2025)
7. Cruz-Romero, M.C., Kerry, J.P.: Packaging of cooked meats and muscle based, convenience-style processed foods. In: Processed Meats, pp. 666–705. Woodhead Publishing (2011)
8. da Costa, T.P., Gillespie, J., Cama-Moncunill, X., Ward, S., Condell, J., Ramanathan, R., Murphy, F.: A systematic review of real-time monitoring technologies and its potential application to reduce food loss and waste: key elements of food supply chains and IoT technologies. Sustainability **15**(1), 614 (2022)
9. Du, D., Xu, M., Wang, J., Gu, S., Zhu, L., Hong, X.: Tracing internal quality and aroma of a red-fleshed kiwifruit during ripening by means of GC-MS and E-nose. RSC Adv. **9**(37), 21164–21174 (2019)
10. Duguma, A.L., Bai, X.: How the internet of things technology improves agricultural efficiency. Artif. Intell. Rev. **58**(2), 63 (2024)
11. Feng, H., Zhang, M., Liu, P., Liu, Y., Zhang, X.: Evaluation of IoT-enabled monitoring and electronic nose spoilage detection for salmon freshness during cold storage. Foods **9**(11), 1579 (2020)
12. Filimon, V., Andrei, M.: Challenges of food packaging materials. Across **5**(3), 5–11 (2022)
13. Formentini, G., Favi, C., Moroni, F., Pirondi, A.: Engineering design in food-packaging industry: the case study of a tuna canning machine. Procedia CIRP **100**, 229–234 (2021)
14. Gahruie, H.H., Parastouei, K., Mokhtarian, M., Rostami, H., Niakousari, M., Mohsenpour, Z.: Application of innovative processing methods for the extraction of bioactive compounds from saffron (Crocus sativus) petals. J. Appl. Res. Med. Aromat. Plants **19**, 100264 (2020)
15. Gamage, A., Thiviya, P., Liyanapathiranage, A., Wasana, M.D., Jayakodi, Y., Bandara, A., et al.: Polysaccharide-based bioplastics: eco-friendly and sustainable solutions for packaging. J. Compos. Sci. **8**(10), 413 (2024)
16. Guo, Z., Guo, C., Sun, L., Zuo, M., Chen, Q., El-Seedi, H.R., Zou, X.: Identification of the apple spoilage causative fungi and prediction of the spoilage degree using electronic nose. J. Food Process Eng. **44**(10), e13816 (2021)
17. Gupta, V.K., Han, S., Xiao, J., Huh, Y.S.: Use of future-oriented smart sensors in food distribution and safety networks. Food Chem. Int. (2025)
18. Harnkarnsujarit, N., Wongphan, P., Chatkitanan, T., Laorenza, Y., Srisa, A.: Bioplastic for sustainable food packaging. In: Sustainable food processing and engineering challenges, pp. 203–277. Academic Press (2021)
19. Heffer, S., Anastasiadi, M., Nychas, G.J., Mohareb, F.: Fusion vs. isolation: evaluating the performance of multi-sensor integration for meat spoilage prediction. Foods **14**(9), 1613 (2025)
20. Igwe, A.N., Eyo-Udo, N.L., Stephen, A.: Technological innovations and their role in enhancing sustainability in food and FMCG supply chains. Int. J. Eng. Inven. **13**(9), 176–188 (2024)
21. Janga Prasanth, M.Y.A.S.P., Kumar, A., Avik, M.: The Role of Microbes in Food Spoilage and Food Safety (2025)
22. Kılcı, O., Koklu, M.: Automated classification of biscuit quality using YOLOv8 models in food industry. Food Anal. Methods 1–15 (2025)
23. Kuchipudi, J.D., Selma, F., Mumtaz, M.A., Syamand, H., Hussein, M.K., Mercado, G.M.: Application of Nanotechnology in Food Safety and Agriculture: Smart Packaging and Delivery Systems for Enhanced Food Security (2024)

24. Kuswandi, B., Restyana, A., Abdullah, A., Heng, L.Y., Ahmad, M.: A novel colorimetric food package label for fish spoilage based on polyaniline film. Food Control **25**(1), 184–189 (2012)
25. Kuswandi, B., Wicaksono, Y., Jayus, Abdullah, A., Heng, L.Y., Ahmad, M.: Smart packaging: sensors for monitoring of food quality and safety. Sens. Instrum. Food Qual. Saf. **5**, 137–146 (2011)
26. Liakos, K.G., Busato, P., Moshou, D., Pearson, S., Bochtis, D.: Machine learning in agriculture: a review. Sensors **18**(8), 2674 (2018)
27. Liang, X., Jia, X., Huang, W., He, X., Li, L., Fan, S., et al.: Real-time grading of defect apples using semantic segmentation combination with a pruned YOLO V4 network. Foods **11**(19), 3150 (2022)
28. Lin, J., Jaiswal, A.K., Jaiswal, S.: A critical review of consumer perception and environmental impacts of bioplastics in sustainable food packaging. Sustainability **17**(4), 1358 (2025)
29. Liu, X., Cheng, S., Liu, H., Hu, S., Zhang, D., Ning, H.: A survey on gas sensing technology. Sensors **12**(7), 9635–9665 (2012)
30. Ma, M., Yang, X., Ying, X., Shi, C., Jia, Z., Jia, B.: Applications of gas sensing in food quality detection: a review. Foods **12**(21), 3966 (2023)
31. Mani, R., Kumar, J.V., Murugesan, B., Alaguthevar, R., Rhim, J.W.: Smart sensors in food packaging: sensor technology for real-time food safety and quality monitoring. J. Food Process Eng. **48**(4), e70120 (2025)
32. Nadeem, H., Zaid, M., Sajid, M., Jameel, N., Shahid, E., Bukhari, S., Saleem, A., Maqbool, M., Bashir, S., Toqeer, S.: Spore Forming Bacteria-A Major Cause of Food Spoilage and Outbreaks: A Comprehensive Review (2025). https://doi.org/10.47278/book.HH/2025.94
33. Ncube, L.K., Ude, A.U., Ogunmuyiwa, E.N., Zulkifli, R., Beas, I.N.: An overview of plastic waste generation and management in food packaging industries. Recycling **6**(1), 12 (2021)
34. Opara, I.K., Divine, D.C., Silue, Y., Opara, U.L., Okolie, J.A., Fawole, O.A.: Application of machine learning in predicting fruit waste in a South African fresh produce wholesale market. J. Agric. Food Res. 102062 (2025)
35. Pandey, V.K., Srivastava, S., Dash, K.K., Singh, R., Mukarram, S.A., Kovács, B., Harsányi, E.: Machine learning algorithms and fundamentals as emerging safety tools in preservation of fruits and vegetables: a review. Processes **11**(6), 1720 (2023). https://doi.org/10.3390/pr11061720
36. Prakash, J., Kumari, U.: IoT Advancements in Ensuring Food Safety: A Comprehensive Analysis in the Indian Context (2024)
37. Revelou, P.K., Tsakali, E., Batrinou, A., Strati, I.F.: Applications of machine learning in food safety and HACCP monitoring of animal-source foods. Foods **14**(6), 922 (2025)
38. Risch, S.J.: Food packaging history and innovations. J. Agric. Food Chem. **57**(18), 8089–8092 (2009)
39. Sathiamurthy, J., Geetha, R.V., Gopal, R.K.: Production and evaluation of a seaweed-based bioplastic sheet for food packaging. Uttar Pradesh J. Zool. **45**(14), 108–115 (2024)
40. Sharma, S., Gahlawat, V.K., Rahul, K., Mor, R.S., Malik, M.: Sustainable innovations in the food industry through artificial intelligence and big data analytics. Logistics **5**(4), 66 (2021)
41. Shine, P., Murphy, M.D.: Over 20 years of machine learning applications on dairy farms: a comprehensive mapping study. Sensors **22**(1), 52 (2021)
42. Siciliano, S., Lopresto, C.G., Lamonaca, F.: From traditional packaging to smart bio-packaging for food safety: a review. Euro-Mediterr. J. Environ. Integr. 1–16 (2024)
43. Sobhan, A., Hossain, A., Wei, L., Muthukumarappan, K., Ahmed, M.: IoT-enabled biosensors in food packaging: a breakthrough in food safety for monitoring risks in real time. Foods **14**(8), 1403 (2025)
44. Sonwani, E., Bansal, U., Alroobaea, R., Baqasah, A.M., Hedabou, M.: An artificial intelligence approach toward food spoilage detection and analysis. Front. Public Health **9**, 816226 (2022)
45. Surya, V., Senthilselvi, A.: A qualitative analysis of the machine learning methods in food adultery: a focus on milk adulteration detection. J. Adv. Res. Dyn. Control Syst. **12**(7), 543–551 (2020)

46. Taneepanichskul, N., Hailes, H. C., Miodownik, M., & Hailes, H. C. (2025). Using hyperspectral imaging and machine learning to identify food contaminated compostable and recyclable plastics. *UCL Open Environment*, *7*(1).
47. Thul, P P., Jagtap, D.N., Gosavi, R.A., Khatal, M.M., Labade, P., Sawant, A.A.: Biotechnological Innovations in Packaging and Sensors in the Food Processing (2025)
48. Umayanga, U.G.E., Chethana, U.G.K.: Beyond Plastics: Exploring Eco-Friendly Alternatives in Food Packaging-A Review of Both Biodegradable and Bio-based Bioplastic Materials (2024)
49. Valderas, D., Villa-González, F., Bhattacharyya, R., Sarma, S.: Time–temperature excursion monitoring using chipless RFID tags and organic oils. IEEE Sens. J. **23**(17), 19028–19036 (2023)
50. Want, R.: An introduction to RFID technology. IEEE Pervasive Comput. **5**(1), 25–33 (2006)
51. Witjaksono, G., Saeed Rabih, A.A., Yahya, N.B., Alva, S.: IOT for agriculture: food quality and safety. In: IOP Conference Series: Materials Science and Engineering, vol. 343, p. 012023. IOP Publishing (2018)
52. Witjaksono, G., Saputra, I., Latief, M., Jaswir, I., Akmeliawati, R., Rabih, A.A.S.: Non-halal biomarkers identification based on Fourier transform infrared spectroscopy (FTIR) and gas chromatography-time of flight mass spectroscopy (GC-TOF MS) techniques. In: EPJ Web of Conferences, vol. 162, p. 01007. EDP Sciences (2017)
53. Wyrwa, J., Barska, A.: Innovations in the food packaging market: active packaging. Eur. Food Res. Technol. **243**, 1681–1692 (2017)
54. Yang, H., Jiao, W., Zouyi, L., Diao, H., Xia, S.: Artificial intelligence in the food industry: innovations and applications. Discov. Artif. Intell. **5**(1), 1–34 (2025)
55. Yang, X., Scharff, R.: Foodborne illnesses from leafy greens in the United States: attribution, burden, and cost. J. Food Prot. **87**(6), 100275 (2024). https://doi.org/10.1016/j.jfp.2024.100275
56. Zhai, X., Li, Z., Shi, J., Huang, X., Sun, Z., Zhang, D.I., et al.: A colorimetric hydrogen sulfide sensor based on gellan gum-silver nanoparticles bionanocomposite for monitoring of meat spoilage in intelligent packaging. Food Chem. **290**, 135–143 (2019)
57. Zuo, J., Feng, J., Gameiro, M.G., Tian, Y., Liang, J., Wang, Y., et al.: RFID-based sensing in smart packaging for food applications: a review. Futur. Foods **6**, 100198 (2022)

Real-World Applications of AI-Designed Polymers

Rohit Verma and Tejendra Kumar Gupta

Abstract This chapter explores the transformative impact of artificial intelligence (AI) on the discovery, design, and deployment of advanced polymeric materials in real-world applications. By leveraging machine learning and data-driven approaches, researchers have accelerated the identification of novel polymers with tailored properties, addressing challenges that traditional experimental strategies face in navigating vast chemical design spaces. The chapter examines state-of-the-art AI techniques enabling the rapid prediction of polymer performance, virtual screening, and inverse molecular design, supported by examples in sectors such as energy storage, biomedical devices, sustainable packaging, and environmental remediation. Key considerations, such as the interpretability of machine learning models, the integration of experimental synthesis workflows, and issues of processability and scalability, are addressed to bridge the gap between computational discovery and industrial application. By highlighting recent breakthroughs and ongoing challenges, the chapter underscores AI's growing role in engineering next-generation polymers that meet sustainability, performance, and cost-effectiveness criteria for diverse technological domains.

Keywords Artificial intelligence in polymer science · Sustainable polymers · High-throughput screening · Industrial applications of AI-designed polymers

1 Introduction

Polymers are the backbone of countless modern technologies, from packaging materials to aerospace composites and biomedical devices. Traditionally, the discovery and design of new polymers have relied on time-consuming experimental trial-and-error methods or computational techniques with limited predictive power [1]. However, the convergence of artificial intelligence (AI) with polymer science has

R. Verma (✉) · T. K. Gupta
Amity Institute of Applied Sciences, Amity University Uttar Pradesh, Noida, India
e-mail: rverma85@amity.edu

B. Sharma et al. (eds.), *AI-Driven Polymer Design*, Engineering Materials,
https://doi.org/10.1007/978-981-95-8353-9_7

sparked a transformative shift, enabling the rational design of polymers with tailored properties for specific real-world applications Chen et al. [19]; Kim et al. [20].

AI-designed polymers harness advanced machine learning (ML) algorithms to predict structure–property relationships, optimize monomer sequences, and suggest novel polymer architectures far beyond what intuition or classical approaches alone could achieve Wu et al. [21]. By training models on existing experimental datasets, AI systems can identify hidden correlations between chemical structures and macroscopic properties—such as mechanical strength, thermal stability, or biodegradability—dramatically accelerating the materials discovery process Jha et al. [22]. Generative models, including variational autoencoders and generative adversarial networks, have further enabled the in-silico creation of entirely new polymer candidates optimized for complex property targets (Sanchez-Lengeling and Aspuru-Guzik [23].

These developments are no longer confined to academic curiosity; AI-designed polymers have begun to make tangible impacts across diverse industries. For instance, in electronics, AI-driven optimization has led to flexible polymers with unprecedented conductivity and stretchability for next-generation wearable devices Kim et al. [20]. In environmental applications, machine learning-assisted polymer design has produced advanced membranes for water purification and carbon capture, offering potential solutions to global sustainability challenges Li et al. [24]. The healthcare sector has also witnessed breakthroughs: AI-guided discovery of biodegradable and biocompatible polymers is revolutionizing drug delivery systems and regenerative medicine scaffolds, with implications for patient-specific therapies (Kumar et al. 2020).

Despite these promising advances, challenges remain. Reliable datasets for polymer properties are often scarce or inconsistent, limiting the predictive accuracy of AI models. Moreover, the interpretability of deep learning algorithms is still an open question, raising concerns about the trustworthiness of AI-generated predictions in safety–critical applications like medical implants or food-contact materials. Addressing these issues will require interdisciplinary collaboration between chemists, materials scientists, computer scientists, and engineers, as well as the development of standardized, high-quality polymer datasets.

This chapter explores the current state of AI-designed polymers, highlighting real-world examples that demonstrate the practical potential of these materials across multiple sectors. We review key methods in AI-driven polymer design, discuss their strengths and limitations, and present case studies illustrating how AI has enabled rapid innovation in fields ranging from sustainable packaging to biomedical engineering. By providing a comprehensive overview, this chapter aims to serve as a valuable resource for researchers, practitioners, and students interested in the emerging intersection of artificial intelligence and polymer science.

2 AI in Polymer Design

Tools and Techniques AI approaches such as supervised learning, unsupervised learning, reinforcement learning, and generative models are employed in polymer science. Tools like deep neural networks (DNNs), support vector machines (SVMs), and Gaussian processes (GPs) are commonly used to model complex relationships between molecular structure and macroscopic properties. Generative models, including variational autoencoders (VAEs) and generative adversarial networks (GANs), are particularly useful for creating novel polymer structures with desired characteristics. Databases like Polymer Genome, ChemSpider, and Materials Project provide essential data for training AI models. Moreover, cheminformatics tools and molecular descriptors help quantify polymer structures, aiding in accurate predictions.

Polymers are essential materials—ranging from plastics to elastomers—used across industries such as packaging, construction, electronics, and biotechnology. Designing polymers with precise thermal, mechanical, or functional properties has historically been a slow, experimental, and expensive endeavor due to the vast chemical design space and complex structure–property relations.

Enter Artificial Intelligence (AI) and Machine Learning (ML): data-driven approaches capable of handling high-dimensional, non-linear relationships and learning from both experimental and simulated datasets. These technologies are increasingly being tapped to accelerate polymer design, property prediction, and synthesis planning, shifting the field from empirical trial-and-error to predictive, efficient, and scalable approaches.

3 Foundations: Data, Descriptors, and Databases

Polymer Databases and Data Acquisition

PoLyInfo and MSA are key polymer databases used for collecting permeability and structural data, such as in the development of ML models to predict gas separation performance (Yang et al. 2022). Data scarcity and fragmentation remain significant challenges; high-quality, standardized datasets are rare, inhibiting robust ML model building (Long et al. 2025; Karuppusamy et al. 2025).

Molecular Descriptors and Representations

Machine learning requires encoding polymer structures numerically:

Molecular descriptors derived from polymer repeating units (e.g., atom counts, ring presence) and fingerprints like Morgan fingerprints (MFF) are commonly used to represent structure (Long et al. 2025; Karuppusamy et al. 2025).

The review by Zhao et al. [2] in Polymer Chemistry provides an extensive overview of descriptor types (constitutional, topological, etc.) and their use in ML models

(Karuppusamy et al. 2025). Table 1 depicts representative AI-designed polymers and their impact.

Supervised Learning for Property Prediction

Random Forests (RFs) and Deep Neural Networks (DNNs) have been used to predict gas permeabilities across polymers, including strategies to handle incomplete datasets via imputation (MICE algorithm) (Long et al. 2025). Deep learning models, e.g., DNNs with architectures optimized for polymer tasks, show strong predictive performance (Long et al. 2025).

Graph-Based and Deep Representations

Graph Convolutional Neural Networks (GCNNs) can predict dielectric constant and bandgap with accuracy matching DFT, without handcrafted descriptors [3].

polyBERT, a transformer-based chemical language model treating polymer SMILES-like strings as "language," provides ultrafast, scalable fingerprinting with high predictive accuracy [3]. Table 2 provides a list of polymers designed by using AI.

Table 1 Representative AI-designed polymers and their impact

Polymer/class	Year	References
Polynorbornene/polyimide for capacitors	2024	Nature Communications, Ramprasad et al.
Sustainable custom plastics	2025	Science Advances, UChicago, de Pablo et al.
Membranes for CO_2 separation	2023	Nature Communications (inverse design)
ML-identified recyclable polymers	2024	Nature Reviews Materials, Ramprasad et al.
Polymer-based organic batteries	2016	*Chemical Review*, Muench et al.
Advanced polymeric dielectrics	2016	*Progress in Material Sciences*, Huan et al.

Table 2 Polymers designed by using AI

Polymer/class	Year	Reference (primary literature/ source)	Application/notes
High energy density polyimide/ polynorbornene	2024	Nature Communications (Georgia Tech, Ramprasad et al.)	Electrostatic energy storage, aerospace, high stability
Custom sustainable plastics	2025	Science Advances (UChicago, Juan de Pablo et al.)	Environmentally degradable, custom mechanical properties
Polymer membranes for CO_2 capture	2023	Nature Communications (inverse design)	Membranes for carbon capture/separation
Accelerated design via ML-assisted molecular screening	2025	Accounts of Materials Research (ML for molecular design)	General polymer discovery, rapid property optimization
ML-identified recyclable plastics	2024	Nature Reviews Materials (Ramprasad et al.)	Sustainable, circular economy-focused

Table 3 Selected real-world applications for AI-designed polymers

Application	AI-designed polymer	Key properties	References
Energy storage	Polynorbornene/ polyimide	High dielectric and breakdown strength	[4]
Carbon capture	Novel polymer membranes	Selectivity, permeability, robustness	[5]
Biomedicine	Tailored hydrogels and scaffolds	Tunable mechanics, biocompatibility	[6, 7]
Packaging	Biodegradable custom plastics	Degradable, barrier property tailored	Taofeek et al. (2022)

Multi-Task Learning and Interpretability

Multi-task learning architectures trained on thousands of polymers (across 36 properties) allow models to leverage correlations between different properties, improving performance and offering interpretable insights for design [3].

Inverse Design and Optimization

Inverse design via ML enables specifying a target property (e.g., cloud-point temperature) and searching for polymer structures to achieve it. Example: gradient boosting decision trees achieved ~4 °C RMSE in cloud-point prediction, later validated experimentally [3].

Studies advocate combining high-throughput experimentation, simulations, and ML to create integrated design platforms enabling inverse design (Kumar et al. 2019). Table 3 shows selected real-world applications for AI-designed polymers.

Active Learning and Autonomous Experimentation

A Learn–Design–Build–Test cycle using Gaussian Process Regression (GPR) and Bayesian optimization on robotic platforms efficiently identified high-reactivity polymer–protein hybrids (PPHs), integrating ML with automated synthesis and testing [8].

4 Benefits of AI-Driven Polymer Design

- **Speed and efficiency**: Rapid screening of vast chemical spaces, reducing experimental cycles.
- **Cost reduction**: Decreases material waste and experimental failures.
- **Novelty generation**: Captures unexplored candidates beyond human intuition.
- **Data integration**: Combines simulation, experimental, and descriptor-driven insights.

- **Interpretability (in some models)**: Multi-task and explainable approaches enable understanding of feature–property relations [9].

5 Real-World Applications

5.1 Healthcare and Biomedical Applications

AI-designed polymers have made significant strides in healthcare. Smart drug delivery systems using stimuli-responsive polymers can release therapeutics at targeted sites in response to pH, temperature, or enzymes. For example, AI models have helped design hydrogels for controlled drug release with tunable degradation rates and mechanical properties [10]. In tissue engineering, AI-designed biodegradable scaffolds support cell growth and differentiation. By predicting optimal compositions and architectures, AI enhances the performance of polymer scaffolds for regenerative medicine [11].

5.2 Energy Storage and Conversion

Polymers play a crucial role in energy devices, such as batteries, fuel cells, and solar cells. AI-designed polymer electrolytes offer improved ionic conductivity and thermal stability. For instance, AI has guided the discovery of polymer electrolytes for lithium-ion batteries with superior cycle life and safety [12]. In organic photovoltaics (OPVs), AI-assisted design has led to high-efficiency donor–acceptor polymers with optimized light absorption and charge mobility. These advancements contribute to cost-effective and flexible solar energy solutions [13]. Table 4 collets key AI tools and platforms in polymer design.

Table 4 Key AI tools and platforms in polymer design

Tool/platform	Functionality	Industrial adoption evidence	References
Polymer informatics	Property prediction/ modeling	Matmerize Inc., Toyota, GE	[6, 7]
ML-guided simulation	Virtual screening, optimization	Dow, BASF, startup accelerators	[16]
Automated synthesis	Robotic experimentation	Next-gen labs (university, private)	[17]
Generative models	De novo structure generation	Early trial phase	[18]

5.3 *Electronics and Optoelectronics*

Flexible electronics and wearable devices benefit from AI-designed conductive and semiconductive polymers. Machine learning models optimize properties such as conductivity, transparency, and mechanical flexibility. Polymers with tailored bandgaps and charge transport characteristics enable high-performance organic transistors, LEDs, and sensors [14]. Additionally, AI helps design polymers for printed electronics and photonic devices, enhancing their functionality and integration in complex systems.

5.4 *Environmental Sustainability*

AI is instrumental in developing sustainable polymers to reduce plastic waste and environmental impact. Biodegradable and recyclable polymers are designed with lifecycle considerations. AI models predict degradation pathways, mechanical resilience, and recyclability, facilitating the development of environmentally friendly materials [15]. Furthermore, AI aids in upcycling strategies by identifying compatible additives and blending conditions to enhance the properties of recycled polymers. Different types of polymers and their applications are shown in Fig. 1.

6 Industrial Implementation and Commercialization

Several companies and research institutions are integrating AI into polymer R&D. For example, IBM's Materials Discovery program uses AI to identify polymers for water purification and carbon capture. BASF and Dow Chemical have invested in AI-driven materials platforms to accelerate product development. Startups like Kebotix and Citrine Informatics employ AI to automate polymer discovery, reducing development time and cost. These efforts are bridging the gap between laboratory research and commercial production.

7 Challenges and Future Directions

Despite promising developments, challenges remain in data quality, model interpretability, and experimental validation. Incomplete or biased datasets can lead to inaccurate predictions. Interpreting complex AI models is crucial for gaining scientific insights and ensuring reliability. Integrating AI with high-throughput experimentation and robotics can enhance validation and feedback loops. Moreover, establishing standardized data formats and open-access repositories will improve

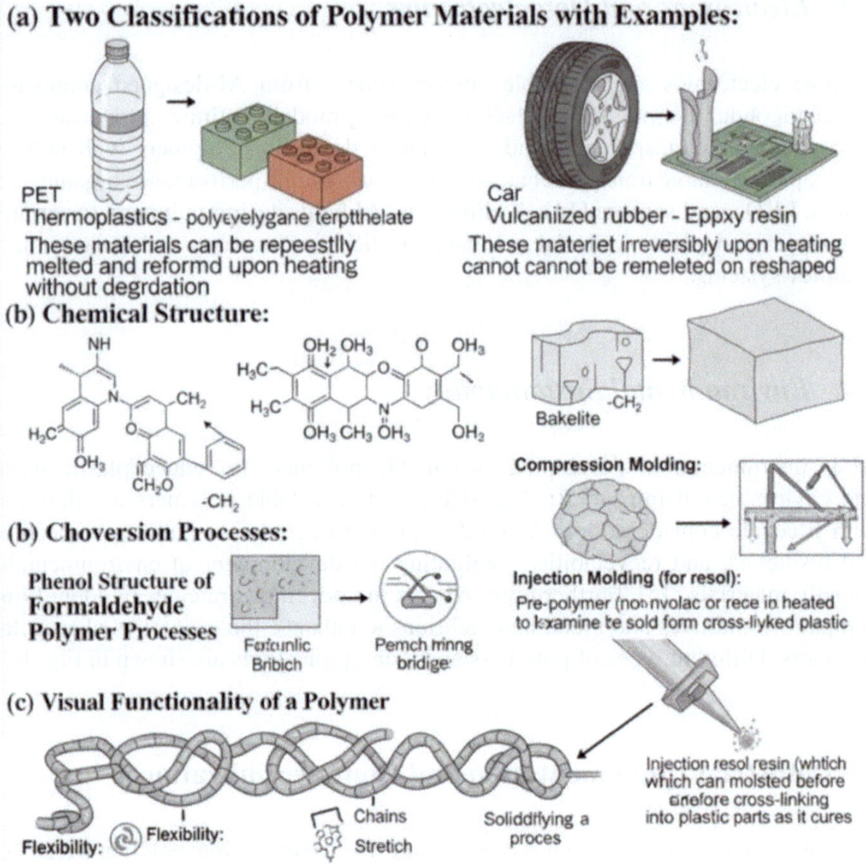

Fig. 1 Different types of polymers and their applications

model training and reproducibility. Future research should focus on hybrid models combining physics-based simulations with AI, enabling more accurate and generalizable predictions. Ethical considerations, including data privacy and algorithmic bias, should also be addressed.

8 Conclusion

AI-designed polymers represent a transformative leap in materials science, enabling faster, smarter, and more sustainable innovations. By harnessing machine learning and data-driven methodologies, researchers are developing advanced polymers tailored for specific real-world applications. As AI continues to evolve, its integration with polymer science will unlock new frontiers in healthcare, energy, electronics, and environmental sustainability.

References

1. Tran, H., Gurnani, R., Kim, C., et al.: Design of functional and sustainable polymers assisted by artificial intelligence. Nat. Rev. Mater. **9**, 866–886 (2024). https://doi.org/10.1038/s41578-024-00708-8
2. Zhao et al.: A review on the application of molecular descriptors and machine learning in polymer design. Polym. Chem. (2023)
3. Kumar et al.: Cloud-point engineering via ML inverse design. arXiv (2018)
4. Kanezashi, M., Tomarino, Y., Nagasawa, H., Tsuru, T.: Tailoring the molecular sieving properties and thermal stability of carbonized membranes containing polyhedral oligomeric silsesquioxane (POSS)-polyimide via the introduction of norbornene. J. Membr. Sci. **582**, 59–69 (2019). ISSN: 0376-7388. https://doi.org/10.1016/j.memsci.2019.04.003. https://www.sciencedirect.com/science/article/pii/S037673881930256X
5. Ma, W., Lin, Y., Tang, Y.: Novel polymeric membranes preparation and membrane process. Separations **9**(9), 253 (2022). https://doi.org/10.3390/separations9090253
6. Chen, S.-Y., Cho, Y.-C., Yang, T.-S., Ou, K.-L., Lan, W.-C., Huang, B.-H., Saito, T., Tsai, C.-H., Peng, B.-Y., Chuo, Y.-C., et al.: A tailored biomimetic hydrogel as potential bioink to print a cell scaffold for tissue engineering applications: printability and cell viability evaluation. Appl. Sci. **11**(2), 829 (2021). https://doi.org/10.3390/app11020829
7. Chen, L., Pilania, G., Batra, R., Huan, T.D., Kim, C., Kuenneth, C., Ramprasad, R.: Polymer informatics: current status and critical next steps. Mater. Sci. Eng. R: Rep. 144, 100595 (2021). ISSN: 0927-796X. https://doi.org/10.1016/j.mser.2020.100595. https://www.sciencedirect.com/science/article/pii/S0927796X2030053X
8. Tamasi et al.: Learn–design–build–test cycle for polymer–protein hybrids. Adv. Mater. (2022)
9. Künneth et al.: Multi-task learning in polymer informatics. arXiv (2020)
10. Zhou, Y., et al.: Smart polymer hydrogels for controlled drug delivery: design and applications guided by machine learning. ACS Appl. Mater. Interfaces **12**(4), 4506–4517 (2020)
11. Lee, H., et al.: AI-driven scaffold design for tissue engineering. Biomater. Sci. **9**(18), 6032–6043 (2021)
12. Chen, J., et al.: AI-assisted discovery of polymer electrolytes for lithium-ion batteries. Nat. Mater. **21**(5), 435–443 (2022)
13. Sun, Y., et al.: Artificial intelligence in organic photovoltaics: materials discovery and device optimization. Energy Environ. Sci. **16**(3), 689–703 (2023)
14. Kumar, A., et al.: Machine learning for polymer design in flexible electronics. Adv. Mater. **33**(14), 2008055 (2021)
15. Wang, T., et al.: Data-driven design of biodegradable polymers for sustainable applications. Green Chem. **23**(6), 2304–2315 (2021)
16. Yu, Y., Yao, H., Liu, Y.: Structural dynamics simulation using a novel physics-guided machine learning method. Eng. Appl. Artif. Intell. **96**, 103947 (2020). ISSN: 0952-1976. https://doi.org/10.1016/j.engappai.2020.103947. https://www.sciencedirect.com/science/article/pii/S0952197620302670
17. Yu, L., Chen, B., Li, Z., Su, Y., Jiang, X., Han, Z., Zhou, Y., Yan, D., Zhu, X., Dong, R.: Computer-aided automated flow chemical synthesis of polymers. Giant **18**, 100252 (2024). ISSN: 2666-5425. https://doi.org/10.1016/j.giant.2024.100252. https://www.sciencedirect.com/science/article/pii/S2666542524000171
18. Yue, T., Tao, L., Varshney, V., Li, Y.: Benchmarking study of deep generative models for inverse polymer design. Digit. Discov. **4**(4), 910–926 (2025). ISSN: 2635-098X. https://doi.org/10.1039/d4dd00395k. https://www.sciencedirect.com/science/article/pii/S2635098X25000518

19. Chen, C.K, Huang, P.K, Law, W.C, Chu, C.H, Chen, N.T, Lo, L.W.: Biodegradable polymers for gene-delivery applications. Int J Nanomedicine **30**(15), 2131–2150 (2020) PMID: 32280211; PMCID: PMC7125329. https://doi.org/10.2147/IJN.S222419
20. Kim, J., et al.: Fracture, fatigue, and friction of polymers in which entanglements greatly outnumber cross-links. Science **374**, 212–216 (2021). https://doi.org/10.1126/science.abg632
21. Wu, L., Shi, X., Z.-S, Wu.: Recent advancements and perspectives of biodegradable polymers for supercapacitors. Adv. Funct. Mater. **33**, 2211454 (2023). https://doi.org/10.1002/adfm.202211454
22. Jha, S., Akula, B., Enyioma, H., Novak, M., Amin, V., Liang, H.: Biodegradable biobased polymers: A review of the state of the art, challenges, and future directions. Polymers **16**, 2262 (2024). https://doi.org/10.3390/polym16162262
23. Sanchez-Lengeling, B., Aspuru-Guzik, A.: Inverse molecular design using machine learning: Generative models for matter engineering. Science **361**, 360–365 (2018). https://doi.org/10.1126/science.aat2663
24. Li, Huiling., Huo, Yan-Gao., He, Xi., Yao, Liping., Zhang, Hao., Yiqiang, Cui., Xiao, Huijuan., Xie, Wenxiu., Dejiu, Zhang., Wang, Yue., Zhang, Shu., Tu, Haixia., Cheng, Yiwei., Guo, Yueshuai., Cao, Xintao., Zhu, Yunfei., Jiang, Tao., Guo, Xuejiang., Qin, Yan., Sha, Jiahao. Li et al-2022-Nature. Nature (2022). 2022. 1. 10.1038/s41586-022-05508-0

Machine Learning Based Innovation in Polymers for Bio Medical and Energy Storage: Enhancing Safety and Sustainability

Vijay Chaudhary, Devang Rana, Chanchal Ahlawat, and Bhasha Sharma

Abstract The article discusses how the use of artificial intelligence (AI) can be transformative in all aspects of design, development, and deployment of advanced polymeric materials. Through the incorporation of AI methods (machine learning, deep learning, and reinforcement learning), scientists are transforming conventional polymer science by means of fastened discovery, predictive modelling and streamlining of material characteristics. Various applications of AI-designed polymers into real-life situations are showcased by the work in terms of biomedical, energy storage, and environmental sustainability areas. Examples of case studies are AI-optimized drug delivery systems, scaffold designs in tissue engineering and AI-enabled 4D printing of medical devices. Also, the study emphasizes the advances to lithium-ion batteries, solid polymer electrolytes, and biodegradable polymer design which are based on AI. In spite of its possibilities, the study admits difficulties both in the computational resources and information and educative deficiency, promoting uniform data sharing and transdisciplinary training. The fusion of AI and polymer science does not just imply speeding up the innovation process but it implies that the process will be more sustainable and smarter in terms of designing the future application of materials.

Keywords Biodegradable polymers · Data-driven polymers · Toxicity · Safety

V. Chaudhary (✉) · D. Rana
Department of Mechanical Engineering, Amity School of Engineering and Technology, Amity University Uttar Pradesh, Noida, India
e-mail: vijaychaudhary111@gmail.com

C. Ahlawat
School of Computer Science Engineering & Technology (SCSET), Bennett University, Greater Noida, Uttar Pradesh, India
e-mail: chanchalchaudhary827@gmail.com

B. Sharma
Department of Chemistry, Deen Dayal Upadhyaya College, University of Delhi, New Delhi, India

B. Sharma et al. (eds.), *AI-Driven Polymer Design*, Engineering Materials,
https://doi.org/10.1007/978-981-95-8353-9_8

1 Introduction

Polymers are macro molecules which are made of repeated units or monomers, and they have a wide range of properties and hence are important in different industries like packaging, biomedical and electronics. Artificial intelligence (AI) has helped to improve their versatility by transforming the design and development of novel polymers. With AI methods (including machine and deep learning), it is possible to effectively estimate properties (including physical properties) of polymers and other materials, thus speeding up the discovery of new materials with desired properties. An example is the AI-aided inversed design of recyclable vitrimeric polymers, which goes a long way towards sustainability [1]. Also, the AI models such as syntax-directed variational autoencoders have been used to create polymers that are made to resist extreme conditions including high temperatures and high electric fields. Moreover, predicting the phase behaviour of polymers and the phase change between them using deep learning based data, which is done mainly by dPOLY, has been widely adopted to enhance material development with specialized properties to be used in highly sophisticated applications [2].

Genetic Algorithm is Artificial Intelligence methodology which was envisaged based on the concept of natural evolution. Giro et al. [3] used the method to make some new conducting polymers of better electronic qualities. They used Genetic Algorithm in combination with some other method named Negative Factor Counting (NFC) technique to minimize the relative concentration of monomeric units in polymeric alloys and consequently to make the design of materials efficient. This method enabled them to identify the structures which fit the set requirements such as some electron-gaps values or metallic conductivity. Their methodology has offered an alternative to trial and error methodology in material design and has hopefully led the experimentalists into the creation of new conducting polymers.

Huang et al. [4] created an AI-based approach to formulate high thermal conductivity polymers. They have trained a deep neuron that foresee thermal conductivity based on analyzing the data of 1144 polymers and chemical fingerprints known as Morgan fingerprints with frequency. They applied Shapley Additive Explanations to find out the influence of various substructures on thermal conductivity. With that, they constructed a library of 32 small polymer fragments and employed two types of optimization strategies, Multi-objective Evolutionary Algorithm and Multi-objective Bayesian Optimization to identify polymers that did not just possess large thermal conductivity, but were also simple to synthesize. In a nutshell, they also simulated to evaluate the thermal conductivity of their best polymer among them and how their chain structures influenced thermal conductivity.

In this chapter, the author has discussed practical applications in which AI-designed polymers can be implemented in different sectors with so much potential.

2 AI-Designed Polymers

Polymers designed with AI could be termed as those polymer-based materials that are created with assistance of the AI tools, like deep learning and Machine learning. These artificial intelligence methods will process huge amounts of data and find recurrent trends to enable us to forecast, optimise and synthesize polymers that have customised and specific properties to a unique application (Fig. 1).

Computational models enable AI-based technologies to easily identify new and complexes polymer structures with the necessary electrical, chemical and mechanical properties. They significantly reduce the extravagant environmental consumption of time, finance and other resources as compared to the more traditional experimental mediums. Generative models such as reinforcement learning and deep neural nets allow the easy construction of new polymers with favourable characteristics. With the help of AI, several sustainable and biodegradable polymers have been found, which helps us to work on a better ecology. This resolves a couple of concerns as regards how plastic waste must be disposed. Another significant benefit of AI implementation in polymer research occurs in the possibility of the AI system to optimise the entire process of polymer synthesis with the help of a range of optimisation mechanisms prompted by AI. These methods are essential in order to determine the most optimal reaction conditions, to economize, and increase reproducibility. Gomez-Bombarelli et al. [5] used artificial intelligence (AI) to design molecules better by enhancing a data-driven, continuous representation of molecules. They deployed a Variational Autoencoder (VAE) to tract the discrete chemical structures with the aid of a latent

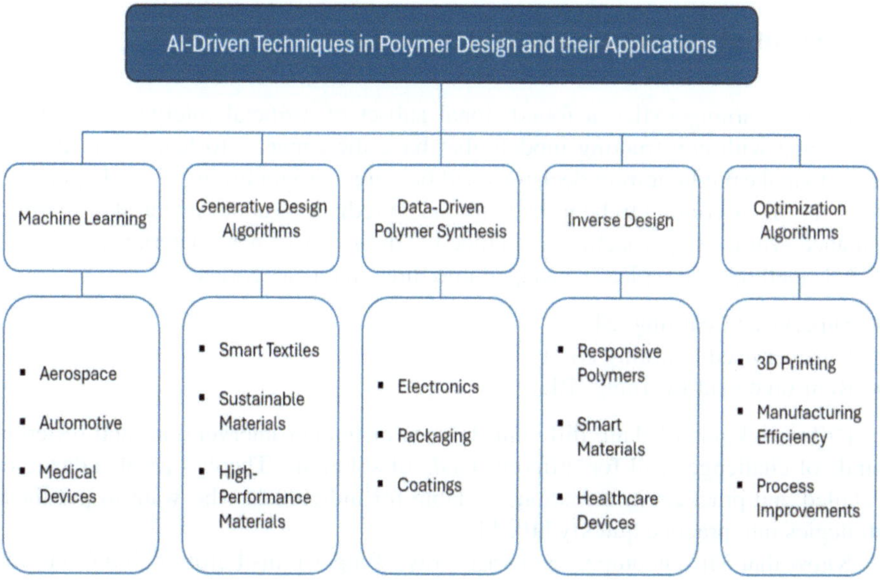

Fig. 1 Applications of AI-driven techniques in polymer design

space to allow productive optimisation and exploration. They could accomplish this by training deep neural networks that comprise a predictor of chemistry property, a decoder of building a molecular composition of a molecule as well as an encoder of converting molecules into consistent vectors. They employed gradient-based optimisation in this latent space to create molecular structures with better properties demonstrating how AI can aid in drug discovery and material development.

Because of the need to encourage innovation and discovery, Jayaraman et al. [6] integrated polymer science with AI and ML. When attempting to build on ML predictions they explored alternative molecular descriptors and developed BigSMILES, a machine-readable chemical structure format. They also deployed AI powered analysis together with high-throughput (HT) experimentation, as a means to optimise the efficiency of experiments, and thereby enable the rapid characterisation, synthesis and prediction of property. ML models were used to automate structure–property mapping, optimise polymer design and manufacture in specific use cases and make sense of complex polymer characterisation data. This combination of AI with molecular models and simulations allowed decreasing the computing costs and increasing the accuracy of the predictions of polymer behaviour.

At the merger of artificial intelligence and polymer informatics are high-performance polymers used in healthcare, electronic and environmentally sustainable packaging. The future of the polymer science will lead to the role of artificial intelligence (AI) becoming even bigger when creating previously visualized materials considered to be impossible to run. This converging is transforming our sustainable development and utilization of future materials, and it is catalyzing innovation.

3 Essential Techniques in Machine Learning

Machine Learning (ML), a foundational subset of artificial intelligence (AI), is concerned with constructing models that have the capacity to learn from data in order to make predictions or decisions without direct programming [7, 8]. In polymer science, ML is coming to be utilized to make predictions about material properties, enhance synthesis approaches, and guide the discovery of new materials [9].

ML methods are typically grouped into three main categories:

- Supervised Learning (SL)
- Unsupervised Learning (UL)
- Reinforcement Learning (RL).

Different kinds of data, different degrees of human intervention, and different kinds of challenges call for different kinds of solutions. There is an abundance of detailed and practical information available for individuals who want to put these strategies into practice quickly [10, 11].

Know that ML encompasses a vast array of algorithms before delving into the specifics of each approach. One of these is deep learning (DL), which is particularly useful when dealing with complicated and big datasets. Machine learning techniques

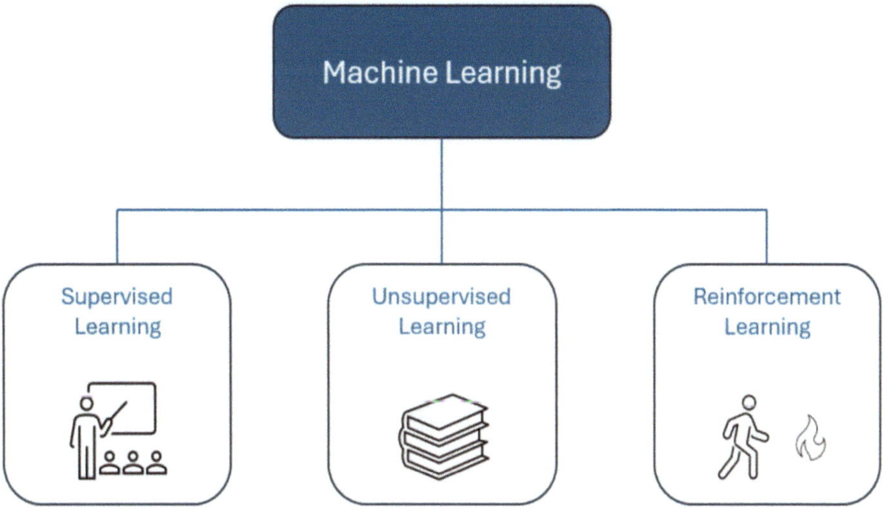

Fig. 2 Machine learning techniques [14]

such as DL, Random Forests, SVMs, PCA, and K-Means Clustering are among the most popular. Materials scientists often employ these techniques because they are ideal for evaluating little datasets, finding patterns in complicated data, and creating models that can be explained. It is critical to select the appropriate method for a given issue [12, 13] (Fig. 2).

Supervised Learning (SL)

One method for training models is supervised learning (SL), which makes use of datasets where the outcomes of each input are known. A classroom setting, with an instructor present to offer guidance and feedback, is a fitting metaphor. Even in SL, the given input–output pairs help the model learn and improve. An SL model, in contrast to a human student, constantly evaluates its own performance and incrementally adjusts its internal parameters until it reaches a trustworthy and accurate prediction point.

Classification and regression are the two main areas where SL is used. Models are taught to classify data in classification issues. As an example, it can be taught to differentiate between biodegradable and non-biodegradable polymers based on their chemical composition and degradation process. With the passage of time, the model can accurately forecast if a new polymer will naturally decay or not. But forecasting continuous values is at the heart of regression difficulties. The prediction of the glass transition temperature, the point at which a polymer changes from a hard, glassy material to a soft, rubbery one, is one such example. An SL model can aid researchers in the development of new materials by accurately predicting the glass transition temperature by studying the correlation between a polymer's molecular structure and its physical characteristics [14] (Fig. 3).

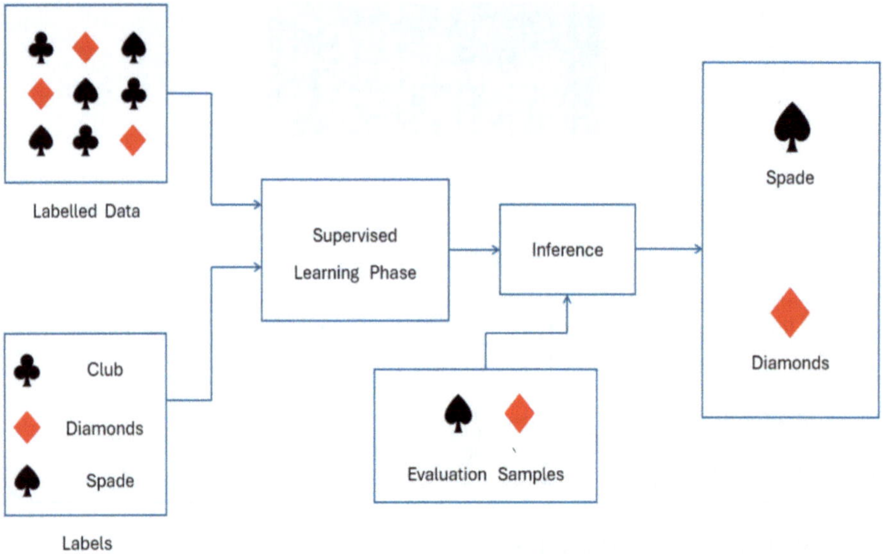

Fig. 3 Supervised learning

Supervised learning has had numerous applications in the sciences of polymer. It contributes in resolving intricate problems made possible by its utilization of vast and detailed databases. The most appropriate example is the prediction of the compatibility of a polymer and a solvent. The model was based on SL to predict these predictions by Chandrasekaran et al. [15], which is much more efficient. The model training dataset contained over 4,500 polymers and 24 different solvents. Tagging of each pair of polymer–solvent combinations was done as per compatibility. To begin with, chemical structures of the solvent and polymer are translated into numerical values that represent molecular properties such as size, polarity and functional groups. The similarity simplification of such numerical descriptors to make it a mathematical entity called lies in a latent space, after which, the neural network is trained to learn the patterns that govern the interactions of solvents with polymers. The model will allow practicing the compatibility of new combinations of polymer and solvent after the training. This SL method achieved accuracy of about 93% compared to more classical heuristic methods to estimate solubility and one is the Hildebrand or Hansen solubility parameter. Such developments are especially useful when it comes to industries in which timely selection of the right solvent could make a major difference in performance of materials and streamlining processes. These are membrane technology, medicine delivery and plastic recycling.

Unsupervised Learning

Aspects Unsupervised learning (UL), a branch of machine learning, stands out due to its capability to discover latent data patterns without labelled data or predicted output. The UL discovers hidden structure in the data in contrast with supervised learning

(SL) which depends upon clear input–output pairs. Consider the kindergarten child who works alone with the collection of materials found in libraries and trying to find repetitive ideas and concepts.

Unsupervised learning (UL) is rather useful when working with complex polymer datasets whose experimental labels are not too numerous or evident. It is really fantastic when it is applied on things:

- ***Clustering***

It refers to the process of grouping polymers that have similar chemical properties or structural characteristics.

- ***Dimensionality Reduction***

It is employed to simplify large and complex datasets into easy ones without compromising the most important information.

Scientists have exploited the UL methods to make sense of complex data sets that involve research in polymer science. Ziolek et al. [16] used UL to study micelles of four-arm and linear block copolymer to ascertain their nanoscale architecture. By using dimensionality reduction they could cluster molecules into similar-shaped groups of micelles sharing a common outer layer (a corona), and they could examine the diversity of structural differences by dimensionality reduction applied to the set of structural differences. Their work provided an illumination of how the self-assembly of micelles can be seen, a factor that is significant to the two disciplines of medication delivery and the research of biomaterials (Fig. 4).

Another interesting example of the application of unsupervised learning demonstrates the analysis of the near-infrared spectra of polyolefins by Sutliff et al. [17]. Since much of the information is superposed, the laborious problem of interpretation of the complex data produced by this type of spectroscopy, by chemical detail,

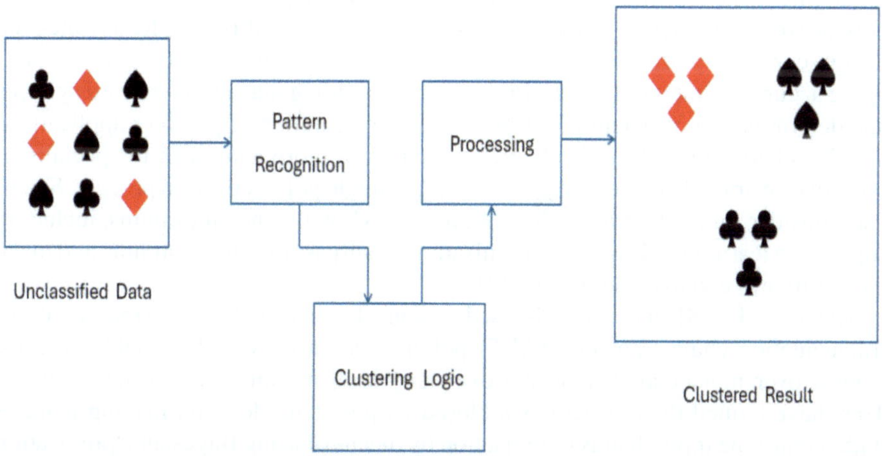

Fig. 4 Illustration of clustering in unsupervised learning

may occasionally crop up. The researchers addressed this difficulty through use of functional principal component analysis (fPCA) which is a statistical tool. This will make the data simplified in that they will have to first focus on the most pertinent variables, it will therefore become easy to identify the major patterns, yet the unnecessary information will be dropped. The team re-interpreted the spectra as continuous representations over all the wavelengths and not a set of discrete data points. Due to this, the method was capable of outlining the prevalent trends and repetitive forms in data. Using this information, we could categorize the polyolefin samples based on their spectral similarities, although we had no categories or labels to make use of. This was a bonus process, since there were apparent correlations between the patterns of the spectra and the chemical composition of the materials, and this enhanced data visualisation and analysis. As a consequence of the same, the inquiry revealed the significant details, which would have been missed through more traditional methods.

Reinforcement Learning

Reinforcement Learning (RL) is one way of machine learning, whereby the models are trained through watching them interact with their environment. The models are bettered depending on the feedback of those interactions. RR enables models to use trial and error to experiment on different tactics and identify the most suitable one. A fitting example would be a child who would learn by feeling something without being explained to that, fire hurts.

RL is best suited to those problems where a sequence of decisions is needed and planning is extensive. However, this is more challenging than the traditional methods of learning. In order to do it, the models have to be adaptive in their nature, able to seek more opportunities at all times and to give their fullest in ensuring that they get the most of the existing ones and working to achieve rewards that may come at some point. It takes much processing power and clever algorithm design to address these problems.

In spite of the fact that RL has many prospects with regards to the enhancement of polymer research, there is still limited use of the concept in this area. RL is time- and energy-consuming because it either demands a huge set of real experiments or developing accurate computer models. Procedures of building managerial reward systems and determination of potential actions under consideration of polymer applications are also challenging. As a result, RL is today in a more natural fit to specialised, objective-restricted purposes, e.g. optimising a single polymerisation process. Scientists consider a more hybrid method of utilizing RL with other algorithms, including supervised learning or Bayesian optimisation, to surpass existing restraints and make the ability more viable and useful [14].

Knox et al. [18] made an AI-based system that can make reversible addition-fragmentation chain transfer (RAFT) polymerisation better. This enables precise control over molecular weight and dispersity with minimum experimental trials. They have termed their system as a closed-loop system, decision making is made within real-time input. It aids optimisation by means of using Bayesian Optimisation (BO) and Thompson Sampling Efficient Multi-Objective Optimisation (TSEMO) technique in particular. The process uses in-real-time feedbacks of nuclear magnetic resonance (NMR) and gel permeation chromatography (GPC) to amend its strategy when it successively conducts a trial with various reaction conditions, compares the results, and reacts accordingly. The system learns by conducting its experiments instead of using already available data. It ensures optimum reaction time and temperature among others in order to optimise monomer conversion and minimal dispersion. To make more effective decisions, the platform creates a model out of what could possibly occur in hypothetical unexplored situations. It then selects the subsequent tests, which exhibit the greatest promise. The logic behind it consists in the fact that in order to optimise simultaneously on multiple goals, the algorithm needs to reach an equilibrium between exploring new reaction scenarios and exploiting the best-understood situations.

4 Biomedical Applications of AI

Artificial intelligence (AI) is revolutionizing biomedical applications in the polymer science field, especially, in drug delivery, and tissue engineering applications. AI assists researchers in creating and optimising systems made of polymers more precisely through the analysis of large data and the prediction of material properties. Consequently, this has been very effective in enhancing efficacy and individualisation of pharmacologic and regenerative medicine therapies.

4.1 AI in Drug Delivery Systems

The drug delivery system utilising polymers was also improved significantly with the application of AI-based techniques and optimised the drug properties such as biodegradability, stability and releasement properties. Machine learning algorithms can be used to anticipate drug-polymer interaction and, subsequently, controlled and sustained drug release and thus enhance therapeutic efficacy and minimize adverse effects [19]. Artificial intelligence (AI) is also useful in creation of nanocarriers through predicting of the interactions of the molecules leading to enhanced effective delivery of drugs and bioavailability of other drugs. Computational models can immediately discover the most appropriate polymer substitutes instead of wasting time and money to use traditional experimental models.

The encapsulation of drugs into hydrogels, micelles, or polymeric nanoparticles and their targeted delivery to tissues is another field where AI has remained very useful in the past few years. Apart from sharing these shortcomings, smart medication delivery systems also employ AI to ensure that the drug is filled in the right place and at the right time by analysing parameters such as degradation rates, pH responsiveness, and temperature sensitivity. It is also essential to employ artificial intelligence (AI) in drug repurposing, as it identifies the most efficient polymer-based drugs personalized according to the individual needs of each patient [20].

4.2 AI-Assisted Scaffold Design in Tissue Engineering

Artificial intelligence (AI) is transforming the tissue engineering landscape as far as the design of the scaffolds is concerned as it is making them bioactive, porous, and sound in terms of their mechanical properties. AI also digs up huge databases of the biological reaction to different materials to find a suitable combination of polymers that may be most effective on a particular tissue. In an attempt to enhance cell adherence, growth, and functioning, deep learning algorithms under the artificial intelligence (AI) could be applied to design 3D-printed scaffolds that approximate the use of the extracellular matrix found in the human body [21].

In the case of repairing the lost bone, artificial intelligence algorithms analyze the properties of biomaterials and provide the most suitable scaffolding designs to stimulate the growth of bones and integrate them. In order to maximise the mechanical and biochemical properties of the scaffold to accommodate the demands of the fragile tissues, artificial intelligence also serves to play an important role in the engineering fields of brain and cardiac tissues. Due to consistent experimentation and application of the reinforcement learning methods, regenerative treatments are becoming more specialist and effective [21].

4.3 AI-Driven Design of Deployable Implants

Jiao et al. [22] explored the implementation of shape memory polymers in an AI-based system of designing an implant. They optimised the microstructures of these implants and made sure that they contained the required mechanical properties, such as bending stiffness, by developing an AI-based inverse design method combining neural networks and evolutionary algorithms. The implants had the potential to reshape according to the temperature change due to the fact that the material used to make the implants was biocompatible polymers including hydrogels composed of poly(ethylene glycol), poly(2-caprolactone); poly(2-caprolactone) (PCL) and poly lactic acid (PLA). Tiny implants could be placed in the body without the need to cut it because of its malleability. These prosthetics then grow to their full and fully functional size when implanted. Two prospective applications had been explored; namely the tracheal stents and also the spinal fusion implants. In order to facilitate insertion and then thermal expansion, the spinal fusion implants have first to be flatted. This would be able to help them support spine better and facilitate bone formation. The shape of the tracheal stent significantly enhances mechanical response; compressed in the delivery, the stent opens up and keeps the airway open. The biocompatibility, practicality, and precision of the surgical and patient-specific standards are followed through implants developed by Jiao et al. [22], which apply a clinically enlightened artificial intelligence design procedure. Implantable devices developed by AI-based deployable thermo-metamaterial, reduced invasiveness of specific patient-centric customised thoroughly functioning medical equipment can be a paradigm shift.

4.4 AI in 4D and 5D Printing for Medical Devices

The use of AI in 4D printing to generate sophisticated and responsive materials was studied by Pugliese et al. [23]. Thanks to artificial intelligence-powered 4D printing, the material can react to changes in mechanical stress, humidity, temperature, and pH. Their remarkable ability to mend themselves, retain their original shape, and even assemble themselves makes them highly useful in the field of medicine. Their research centred on finding new ways to employ computational tools to improve the creation of individualised medical treatments. They zeroed attention on the ways these methods might enhance the production of individualised medical devices, orthoses, and prostheses. Predictive modelling helps scientists find the best materials with the right qualities, reduces the need for trial and error, and may one day allow for direct printing onto human tissue for regenerative medicine and wound healing. Additionally, Pugliese et al. [23] covered two process optimisation approaches, namely open-loop and closed-loop systems, that are employed to enhance fabrication precision and efficiency. In addition, they covered the topic of 5D printing and its new dimension, adaptive computational control, which allows materials to react dynamically in real-time. More efficient, responsive, and patient-specific medical technology is being created by intelligent systems, and their work demonstrates this.

4.5 AI in Biosensing and Artificial Organs

In reference to biosensing, AI comes in handy in the event of producing quality polymer-based biomarker sensors. Diagnosis of diseases such as diabetes and cancer at their beginning stage and offering customised treatment regime to the patients would be impossible without these sensors. AI is also very critical when it comes to development of artificial organs. Bioartificial Lungs, kidneys and bioartificial livers; a synthetic tissue which is created by the researchers utilizing the polymers based on the artificial intelligence. These materials ensure artificial organs are more biocompatible hence perform better when they are inserted into the human body. It could also be applied in regenerative ocular medicine. The artificial intelligence will aid in enhancing polymer-based corneal implants and drugs-dispensing contact lenses. Eye diseases like glaucoma, and corneal injuries can be cured using these technologies causing improvement in eyesight and general well-being. Such innovations indicate how smart polymers are being sought-after in healthcare due to AI.

5 Energy Storage Applications of AI

Electrochemical energy storage system Supercapacitors, fuel cells, and batteries have been under research attention since more than a hundred years. Materials being utilized in the erection of such devices play a major role in their effectiveness, durability, and the capacity of storing energy. Compositions, then, should be devised in a systematic manner, such as electrodes, electrolytes, as well as other catalysts. Conventional methods to synthesize and identify cells to be used in electrochemical devices are very much dependent on trial-and-error applications. Though these methods are tedious and require a lot of work, they are not even reliable. AI was, therefore, resorted to by the scientists in material designing and this gave innovative and efficient resolves to these age-old issues.

5.1 AI in Lithium-Ion Battery Technology

Machine learning is tremendously improving the technology of lithium-ion batteries by improving on the materials that they use in making their electrodes. Some of the most effective and effective methods of predicting cathode material property are as the following:

- Artificial Neural Networks
- Support Vector Machines
- Random Forest
- K-Nearest Neighbours
- Extremely Randomized Trees.

Important for battery performance, these methods use features including density, formation energy, and space group to identify crystal structures and predict electrode voltages. Further, studying lithium-ion intercalation in cathodes relies heavily on methods like Partial Least Squares analysis and ab initio computations [24].

5.2 AI for Solid Polymer Electrolyte Discovery

Bradford et al. [25] produced a chemistry-informed Machine Learning (ML) model that accelerated the generation of Solid Polymer Electrolytes (SPEs) search to energy storage (and mainly in lithium-ion batteries). Other liquid or ceramic electrolytes have better ionic conductivity than current SPE, to which they identified this factor as a limitation. They managed to surpass this by developing ChemArr, a machine learning model. This model forms prediction of the ion conductivity of a range of polymer/salt combinations over a range of temperatures, combining the Arrhenius equation with an MPNN framework. Once trained on more than 12,000 data points of experimental results, the model was used to investigate more than 20,000 potential SPEs, many of which had never previously been experimented on. This is what allowed the scientists to discover materials that were well-designed by AI and would be more effective in real batteries. The work is remarkable in a sense that it does not only provide correct calculations but also utilizes the model to obtain new insights on the structure of molecules and the ionic conductivity that could be used to develop more and more efficient materials in energy storage designing.

5.3 AI-Optimized Capacitor Dielectrics

As the primary devices for storing energy, electrostatic capacitors are indispensable in modern electrical systems. An important determinant of electrostatic capacitor performance, energy density, is the dielectric material. The majority of polymer dielectrics used in this field are either rigid aromatics or flexible polyolefins. It is unusual for these materials to have excellent thermal stability and high energy density at the same time. Gurnani et al. [26] discovered a new family of dielectrics in the polyimide and polynorbornene families by integrating AI with established polymer chemistry and molecular engineering. Some of these recently found materials are very energy dense and thermally stable throughout a large temperature range. Among the newly found dielectrics, one had an energy density at 200 °C of 8.3 Joules per cubic centimetre, eleven times greater than any dielectric used in industry. To make the electrostatic capacitors work better in demanding applications like aerospace without sacrificing environmental sustainability, they looked into ways to improve the PNB and PI families. Particularly in the 85–200 °C temperature range, where viable commercial alternatives are few, they greatly extended the operational range of electrostatic capacitors.

5.4 Machine Learning in Fuel Cell Technology

Fuel cell technology is one of the directions the development of machine learning (ML) is headed. It helps one make better forecasts, interpret complex data and identify trends. Machine learning (ML) can easily cope with the complex physical interactions that occur in fuel cells, unlike the more conventional methods, which rely heavily on such models. The large amount of operational and experimental data which ML is capable of processing enables finding less expensive materials, optimising systems designs, and improving performance due to real-time diagnostics and control [24]. It enables predictive maintenance as well, which optimizes iterative design and testing schedules and increases longevity and robustness of the FC systems.

6 AI in Eco-Polymer Design

With AI, failure analysis and predictions, green materials can be both more efficiently and inexpensively developed as well as made to be less environmentally consumptive. This aids in getting us nearer to a greener future.

6.1 Biodegradable Polymers

The relevance of biodegradable polymers in reducing pollution is growing as they offer more environmentally friendly alternatives to plastics made from petroleum. Chitosan, collagen, cellulose, and starch are some of the natural ingredients that go into making these polymers. Under the influence of microbes, they are engineered to break down enzymatically. Biomedicine, electronics, cosmetics, food packaging, aviation fuels, and textiles are just a few of the many fields that make considerable use of them.

Biodegradability is sensitive to both internal material properties and external environmental factors, making measurement and reporting a significant issue. Loss of weight, carbonyl concentration, and tensile strength are examples of conventional measures that are often imprecise and inconsistent with experimental settings. The establishment of consistent biodegradability databases is made more difficult by this lack of standardisation. In order to overcome this, recent advancements have begun to use AI, specifically ML, for biodegradability prediction. Machine learning techniques provide reliable and scalable biodegradability prediction using big, diverse datasets including chemical structures, environmental conditions, processing histories, and more. The greatest biodegradable alternative to plastics made from petroleum, for instance, was one of over 1.4 million PHA-based polymers screened using deep learning algorithms. An intriguing group of biodegradable polymers is polyhydroxyalkanoates (PHAs), which are produced by microbes. They are excellent candidates to be investigated using AI-based technologies due to their structural diversity and microbial production year-round in appropriate settings. In order to find the best PHA variations for a specific application, predictive ML methods make it easier to map structure–property correlations [27].

The economy continues to face challenges despite these advancements. Still, biopolymer production and recycling are prohibitively expensive compared to plastics made from petroleum, often four to ten times so. Both the complexity of downstream processing and the scalability of production from renewable substrates pose challenges to the commercial viability of this project. To fully utilise AI in biodegradable polymer creation, future studies should focus on improving computational predictions and expanding experimental data sets.

6.2 *Sustainable Management of Waste Polymers*

A sustainable and intelligent approach to reducing landfill use could be to transform polymer waste into high-performance composites. Nevertheless, preserving the material's strength and other mechanical properties when recycling is a major obstacle. In this context, AI becomes useful. Artificial intelligence searches through mountains of experimental data to determine optimal blend ratios and offers suggestions for improving crucial attributes like strength, flexibility, and durability. Reliability in the end product and efficiency in the recycling process are both improved by this.

Reinforced and less fragile high-density polyethylene (HDPE) composites have been successfully manufactured by the integration of evolutionary algorithms with artificial neural networks. Furthermore, advancements in ML have allowed for the creation of cellulose-reinforced recycled polypropylene (PP) composites with mechanical properties comparable to those of virgin PP materials [28].

6.3 *Lifespan Estimation of Sustainable Composites*

Understanding the ageing and degradation processes of sustainable composites is a major challenge. To predict the material's future behaviour, machine learning (ML) is a great alternative to expensive, long-term testing. Machine learning techniques can predict a composite's durability by looking at historical data on material deterioration. In order to predict the behaviour of bio-composites composed of PLA and natural fibres in response to variations in humidity, temperature, and ultraviolet light, algorithms can be trained on such materials. Machine learning (ML) systems have real-world applications in fields like construction and automotive where they may detect signs of wear and strain in materials in real-time, thereby preventing catastrophic failures [28].

7 Challenges to Integrating AI in Polymer Science

Although artificial intelligence (AI) could revolutionise polymer research, there are still many obstacles to overcome before it can be used in real-world applications. Despite a surge in AI-related studies, especially those focussing on machine learning, its implementation in actual labs and factories is still in its infancy. The lack of standardised methods to incorporate ML into current research workflows, computing restrictions, inconsistent or restricted data, and the steep learning curve for scientists unfamiliar with AI are the main causes of this gap. For machine learning to go from being a theoretical concept to a commonly used tool in polymer research, certain obstacles must be overcome.

8 Limited and Fragmented Data

The effectiveness of machine learning (ML) in polymer science is highly dependent on the availability of collaborative platforms and well-structured, high-quality data. Many groups have worked to improve research tools, such as model-sharing repositories, machine-readable polymer formats, and curated databases. Accelerating material discovery and AI-based inventions, these technologies help train and improve ML models. A few examples of resources for polymer research and development are the Materials Project for computational polymer data, Polymer Genome for ML-based property predictions, and the NIST Polymer Database for experimentally validated polymer properties. Although they may not always be compatible with other systems, other domain-specific resources such as P1M and PolyInfo do offer structured data.

Many datasets in polymer science still belong to the "small data" category, with only a few dozen to a few hundred entries—often obtained manually or from literature—regardless of the increasing number of platforms. When it comes to massive datasets, polymer science is still in the dark compared to other industries. This makes ML models, especially deep learning ones, less effective. The sector must prioritise the creation of bigger, better-structured datasets and discover better ways to make data more accessible and shared if it is to progress.

Popular sites like GitHub, Hugging Face, and Zenodo also offer collaborative technologies that can be utilised in polymer science, in addition to repositories that are specifically designed for polymers. For instance, Zenodo is an open-access repository that works well with structured data and reproducible workflows, while Hugging Face has ML models that have already been trained. GitHub makes it easy to manage versions, which improves openness and makes things more reproducible. More complex solutions in polymer science may be modelled after such platforms [14].

But the sector is nevertheless sluggish due to a multitude of issues. Incomplete, inconsistent, or otherwise inaccessible polymer datasets are the norm. They are missing key pieces of information, such the procedure for preparing the substance and measuring its qualities. Open data has already helped artificial intelligence and machine learning flourish in areas like materials science and biology, but polymer science may fall behind if there aren't common standards for organising and sharing this data. Moving forward, it is critical to encourage the inclusion of well-structured datasets in published studies. Making this a regular requirement can have a significant impact on funding organisations and journals. The community's adoption of common formats and standards for the organisation, labelling, and distribution of data is equally important. Machine learning technologies in the field of polymer science can be significantly improved in terms of efficacy and accessibility.

A basic workflow to turn raw data into ML-ready datasets might include:

- Standardizing chemical formats (e.g., using SMILES or BigSMILES),
- Adding metadata (like synthesis or testing methods),
- Cleaning data (removing errors or duplicates), and

Table 1 Depicting global supercomputing initiatives supporting AI-driven polymer research [14]

Sl. No.	Country/ region	Initiative/organization	Supercomputing resource(s)
1	France	Grand Équipement National de Calcul Intensif	Jean Zay
2	European Union	European High-Performance Computing	LUMI, MeluXina
3	United States	Department of Energy (DOE)	Summit, Frontier
4	Japan	RIKEN Centre	Fugaku
5	China	Government-funded AI infrastructure	Facilities in Tianjin and Shenzhen

- Uploading the processed datasets to open platforms like Zenodo, GitHub, or Polymer Genome.

To facilitate reproducibility, data must be digitally accessible, computationally interpretable and version controlled. Researchers should also make their ML workflows and pre-trained models publicly available whenever possible.

To increase accessibility and reusability of data, all datasets must adhere to FAIR principles—Findable, Accessible, Interoperable, and Reusable. The community can foster more trust, better ML outcomes, and speed up innovation in polymer science with a more open and collaborative mindset.

9 Computational Resource Demands

The computing demands of doing AI-based research in polymer science are high. vast computing capacity is required for tasks such as training complex neural networks and processing vast volumes of data from spectroscopy and chemical simulations. Most academic and industrial laboratories do not have the resources to access HPC facilities, which is a common problem. Governments worldwide have stepped in with focused programs to help overcome this obstacle. Table 1 summarises several examples of such attempts.

10 Skill Gaps and Educational Needs

The use of artificial intelligence (AI) in polymer science represents a significant departure for academics whose work is based on experimental or conventional modelling approaches. Integrating AI and ML into basic polymer science curricula is crucial for the field's future success. The complicated chemistry of polymers is beyond the comprehension of most computer scientists, hence it is typically more useful to teach data skills to experts who study polymers.

Despite the growing importance of AI in materials science, the majority of polymer science programs still do not provide students with any kind of formal training in data-driven techniques. Research readiness in this dynamic field is lacking due to the small number of master's programs that provide integrated coursework. Industry, which is experiencing a surge in demand for interdisciplinary competence, and academia are both impacted by this knowledge gap. One solution would be for polymer science courses to incorporate computational tools, data science, and artificial intelligence (AI) straight into the curriculum. Students will be more equipped to apply AI in their future studies if they are exposed to these technologies early on and have actual experience with them.

11 Conclusion

With regards to AI, scientific experts are re-examining the way polymer research is done. It no longer merely makes predictions but is now able to simulate how polymers react to changes in the environment such as pressure, heat and mechanical stress. They are able to experiment with various chemical structures on virtual laboratories and even before any physical lab is created. This has led to simplified, enhanced and less conventional trial and error based development.

Interesting as this tendency is, it becomes still more interesting when one pays attention to the generally wide usage of machine learning, deep learning, and reinforcement learning among other fields of artificial intelligence. Whether scientists are just finding a better battery material, developing a way to make drugs more effective or create smarter medical equipment, artificial intelligence is enabling them to pursue paths that were too complicated or take a time anchor to reach in the past.

Identifying the sustainable materials is one of the most valuable contributions of AI; it promises new horizons of the greener future. The issue of plastic pollution and the quality of the environment are becoming increasingly concerning to more people, so exploring alternatives that would not harm the planet is one of the priorities. With the assistance of artificial intelligence, new biodegradable materials are discovered and optimised which are also effective enough to be used in practice. This is colossal progress in the battle against pollution and the future where there are no fossil fuels.

AI will also assist researchers to optimise such essential properties as the possibility of survival in long-range conditions, electrical conductivity, and heat resistance. It is faster and saves time and resources because it minimises physically run tests.

By enhancing the strength of AI with human knowledge, it functions the best. Being able to quickly work with large data sets and find patterns, AI can only distinguish differences with the help of researchers in their imagination, experience and analytical thoroughness. This partnership is creating a more intelligent, and faster as well as environmentally friendly future of polymer research that is being plotted.

References

1. Doan Tran, H., Kim, C., Chen, L., Chandrasekaran, A., Batra, R., Venkatram, S. et al.: Machine-learning predictions of polymer properties with polymer genome. J. Appl. Phys. **128**(17) (2020)
2. Bhattacharya, D., Patra, T.K.: DPOLY: deep learning of polymer phases and phase transition. Macromolecules **54**(7), 3065–3074 (2021)
3. Giro, R., Cyrillo, M., Galvão, D.S.: Using artificial intelligence methods to design new conducting polymers. Mater. Res. **6**, 523–528 (2003)
4. Huang, X., Zhao, C.Y., Wang, H., Ju, S.: AI-assisted inverse design of sequence-ordered high intrinsic thermal conductivity polymers. Mater. Today Phys. **44**, 101438 (2024)
5. Gómez-Bombarelli, R., Aguilera-Iparraguirre, J., Hirzel, T.D., Duvenaud, D., Maclaurin, D., Blood-Forsythe, M.A., et al.: Automatic chemical design using a data-driven continuous representation of molecules. ACS Cent. Sci. **4**(2), 268–276 (2018)
6. Jayaraman, A., Olsen, B.: Convergence of artificial intelligence, machine learning, chem-informatics, and polymer science in macromolecules. Macromolecules **57**(16), 7685–7688 (2024)
7. Domingos, P.: A few useful things to know about machine learning. Commun. ACM **55**(10), 78–87 (2012)
8. Zdeborová, L.: New tool in the box. Nat. Phys. **13**(5), 420–421 (2017)
9. Aoki, Y., Stephen, W., Tsurimoto, T., Hayashi, Y., Minami, S., Tadamichi, O., Shiratori, K., Yoshida, R.: Multitask machine learning to predict polymer–solvent miscibility using Flory-Huggins interaction parameters. Macromolecules **56**(14), 5446–5456 (2023)
10. Cahill, S.T., Young, J.E., Howe, M., Clark, R., Worrall, A.F., Stewart, M.I.: Assignment of regioisomers using infrared spectroscopy: a Python coding exercise in data processing and machine learning. J. Chem. Educ. **101**(7), 2925–2932 (2024)
11. Lafuente, D., Cohen, B., Fiorini, G., García, A.A., Bringas, M., Morzan, E., Onna, D.: A gentle introduction to machine learning for chemists: an undergraduate workshop using Python notebooks for visualization, data processing, analysis, and modeling. J. Chem. Educ. **98**(9), 2892–2898 (2021)
12. Géron, A.: Hands-on machine learning with Scikit-learn, Keras, and TensorFlow: concepts, tools, and techniques to build intelligent systems. O'Reilly Media, Inc. (2022)
13. Mahesh, B.: Machine learning algorithms—a review. Int. J. Sci. Res. (IJSR) **9**(1), 381–386 (2020)
14. Ferji, K.: Basic concepts and tools of artificial intelligence in polymer science. Polym. Chem. (2025)
15. Chandrasekaran, A., Kim, C., Venkatram, S., Ramprasad, R.: A deep learning solvent-selection paradigm powered by a massive solvent/nonsolvent database for polymers. Macromolecules **53**(12), 4764–4769 (2020)
16. Ziolek, R.M., Smith, P., Pink, D.L., Dreiss, C.A., Lorenz, C.D.: Unsupervised learning unravels the structure of four-arm and linear block copolymer micelles. Macromolecules **54**(8), 3755–3768 (2021)
17. Sutliff, B.P., Goyal, S., Martin, T.B., Beaucage, P.A., Audus, D.J., Orski, S.V.: Correlating near-infrared spectra to bulk properties in polyolefins. Macromolecules **57**(5), 2329–2338 (2024)
18. Knox, S.T., Parkinson, S.J., Wilding, C.Y., Bourne, R.A., Warren, N.J.: Autonomous polymer synthesis delivered by multi-objective closed-loop optimisation. Polym. Chem. **13**(11), 1576–1585 (2022)
19. Vora, L.K., Gholap, A.D., Jetha, K., Thakur, R.R.S., Solanki, H.K., Chavda, V.P.: Artificial intelligence in pharmaceutical technology and drug delivery design. Pharmaceutics **15**(7), 1916 (2023)
20. Serrano, D.R., Luciano, F.C., Anaya, B.J., Ongoren, B., Kara, A., Molina, G., et al.: Artificial intelligence (AI) applications in drug discovery and drug delivery: revolutionizing personalized medicine. Pharmaceutics **16**(10), 1328 (2024)
21. Bagherpour, R., Bagherpour, G., Mohammadi, P.: Application of artificial intelligence in tissue engineering. Tissue Eng. Part B Rev. **31**(1), 31–43 (2025)

22. Jiao, P., Zhang, C., Meng, W., Wang, J., Jang, D., Wu, Z. et al.: Artificial intelligence-guided inverse design of deployable thermo-metamaterial implants. ACS Appl. Mater. Interfaces (2025)
23. Pugliese, R., Regondi, S.: Artificial intelligence-empowered 3D and 4D printing technologies toward smarter biomedical materials and approaches. Polymers **14**(14), 2794 (2022)
24. Liu, X. et al.: Recent advances in artificial intelligence boosting materials design for electrochemical energy storage. Chem. Eng. J. 151625 (2024)
25. Bradford, G. et al.: Chemistry-informed machine learning for polymer electrolyte discovery. ACS Cent. Sci. **9**(2), 206–216 (2023)
26. Gurnani, R., Shukla, S., Kamal, D., Wu, C., Hao, J., Kuenneth, C., et al.: AI-assisted discovery of high-temperature dielectrics for energy storage. Nat. Commun. **15**(1), 6107 (2024)
27. Tran, H., Gurnani, R., Kim, C., Pilania, G., Kwon, H.K., Lively, R.P., Ramprasad, R.: Design of functional and sustainable polymers assisted by artificial intelligence. Nat. Rev. Mater. 1–21 (2024)
28. Karuppusamy, M., Thirumalaisamy, R., Palanisamy, S., Nagamalai, S., Massoud, E.E.S., Ayrilmis, N.: A review of machine learning applications in polymer composites: advancements, challenges, and future prospects. J. Mater. Chem. A (2025)

Future Trends in AI and Polymer Science

Anjali Sharma, Romi Rajput, Vishnu Mittal, and Devkant Sharma

Abstract

Background Polymer science has undergone significant evolution over the years, contributing to numerous innovations in material design, biomedical applications, and sustainability. The integration of Artificial Intelligence (AI) with polymer science represents an emerging frontier that holds great promise in advancing polymer research, development, and application. AI, with its robust data analysis and machine learning capabilities, is poised to revolutionize the design, processing, and performance prediction of polymers. This chapter examines the evolving intersection of AI and polymer science, exploring how this synergy is shaping future trends.

Objective The primary aim of this chapter is to explore the potential of AI in enhancing the study and application of polymers. The focus is on analyzing how AI methodologies, such as machine learning, deep learning, and data mining, can be utilized to optimize the design and properties of polymers for various applications.

Method The methodology involves a comprehensive review of recent literature on the integration of AI technologies in polymer research. Various case studies are discussed, where AI tools have been successfully applied to polymer synthesis, characterizing polymer behaviors, and developing predictive models for polymer properties. The chapter also covers experimental data where AI models were trained to predict polymer properties based on specific molecular structures and processing conditions.

Results The findings demonstrate that AI techniques, such as neural networks and genetic algorithms, have significantly improved the efficiency of polymer design and

A. Sharma (✉)
Guru Gobind Singh College of Pharmacy, Yamunanagar, Haryana, India
e-mail: sharma.sharmaa.anjali@gmail.com

R. Rajput · D. Sharma
Ch. Devilal College of Pharmacy, Jagadhri, Haryana, India

V. Mittal
Adesh College of Pharmacy, Kurukshetra, Haryana, India

B. Sharma et al. (eds.), *AI-Driven Polymer Design*, Engineering Materials,
https://doi.org/10.1007/978-981-95-8353-9_9

processing. AI models have been successfully employed to predict polymer properties, optimize synthesis pathways, and enable the development of high-performance materials. AI has also contributed to the rapid identification of polymer structures with desired functionalities, especially in drug delivery systems, biodegradable polymers, and smart materials.

Conclusion AI and polymer science together represent a powerful combination that will shape the future of material innovation. While challenges remain, including data quality, model interpretability, and computational complexity, the potential benefits of this fusion are immense. Future research should focus on refining AI algorithms for more accurate predictions, addressing sustainability in polymer design, and exploring new applications in biomedical and environmental fields.

Keywords Artificial Intelligence (AI) · Polymer science · Machine learning · Polymer design · Material innovation · Drug delivery systems · Predictive modeling · Sustainability · Smart materials · Biodegradable polymers

1 Introduction

Polymer science, a cornerstone of material science, has profoundly influenced modern society by enabling the development of versatile materials with applications ranging from packaging to biomedical devices. Polymers like polyethylene, polystyrene, and polytetrafluoroethylene (Teflon) have revolutionized industries due to their lightweight nature, flexibility, and resilience [1]. Advances in polymer chemistry have allowed customization of materials at the molecular level, enabling high-performance materials tailored for specific uses such as conductive polymers, shape-memory polymers, and biodegradable plastics [2]. Parallel to this, the field of Artificial Intelligence (AI) has experienced exponential growth. AI, encompassing machine learning (ML), deep learning (DL), and natural language processing (NLP), is transforming how we approach complex problems in research and industry. AI tools can identify patterns, make predictions, and optimize systems without explicit programming abilities that align seamlessly with the challenges of polymer design and development [3]. Merging AI with polymer science marks a paradigm shift in materials research. Traditional trial-and-error methods in polymer development are often time-consuming and expensive. AI offers a data-driven approach to accelerate discovery, reduce waste, and improve predictive capabilities. This fusion opens new avenues for smart materials, real-time process control, and high-throughput material screening [4]. As we venture into this interdisciplinary frontier, understanding the synergy between AI and polymer science is crucial for innovation and sustainability.

2 AI in Polymer Science: A Brief Overview

AI applications in polymer science span a wide range of research areas, including synthesis design, property prediction, and manufacturing optimization. Over the last decade, researchers have increasingly relied on ML algorithms to analyze experimental data, enabling the identification of structure–property relationships that were previously elusive [5]. The influx of large polymer datasets from simulations and experiments has created fertile ground for AI deployment. AI models can now predict how subtle changes in monomer structure or processing conditions affect final polymer properties. Additionally, tools like generative models and reinforcement learning are used to propose novel polymer structures with desired characteristics [6]. Leading institutions and companies are integrating AI into their materials R&D pipelines. For instance, IBM's Materials Discovery Platform and initiatives like the Materials Genome Initiative demonstrate how digital tools can enhance the speed and efficiency of polymer development [7]. As AI becomes more entrenched, a foundational understanding of its core technologies is vital.

3 Core AI Technologies Applied to Polymer Science

3.1 Machine Learning and Deep Learning in Polymer Design

Machine learning algorithms such as support vector machines (SVM), decision trees, and random forests have been employed to map the relationship between polymer structure and mechanical, thermal, or optical properties [8]. Deep learning, a subset of ML, leverages neural networks with multiple layers to extract hierarchical features from complex datasets, enabling breakthroughs in polymer design. Convolutional neural networks (CNNs) and recurrent neural networks (RNNs) are increasingly used to analyse molecular graphs or sequences, identifying patterns not easily seen by humans. These models can be trained to propose polymer structures with improved solubility, thermal stability, or electrical conductivity [9].

3.2 Data Mining and Artificial Neural Networks in Polymer Prediction

Data mining techniques help extract useful insights from vast chemical databases, including PubChem and Materials Project. Artificial neural networks (ANNs), inspired by biological neural systems, learn from input–output data pairs to predict polymer properties. ANNs have been successfully applied to model polymer glass transition temperatures, viscosity behaviour, and biodegradability [10]. Combining these AI models with quantum chemistry or molecular dynamics simulations enables a hybrid approach that improves accuracy and reduces computational cost [11].

3.3 AI-Based Optimization Algorithms for Polymer Synthesis

Genetic algorithms (GAs), particle swarm optimization (PSO), and Bayesian optimization are AI tools used to optimize polymer synthesis parameters such as monomer ratios, catalysts, temperature, and reaction time. These algorithms search vast chemical spaces to identify synthesis routes that maximize desired outcomes [12]. In recent studies, reinforcement learning has also been applied in synthetic planning, guiding the selection of optimal reagents and reaction pathways for polymer formation [13].

4 AI for Predicting Polymer Properties

4.1 Role of AI in Property Prediction

Predicting polymer properties based on molecular or structural information is a critical challenge. Traditional methods, relying on empirical models and simulations, are limited in speed and scalability. AI, particularly ML and DL models, offers a promising alternative for accurate and rapid prediction of properties like tensile strength, dielectric constant, and biodegradability [14]. For instance, a neural network trained on polymer composition data can predict thermal behaviour under specific stress conditions. Similarly, models trained on historical data can estimate barrier properties critical in food packaging or drug delivery systems [15].

4.2 Challenges and Limitations in Property Prediction

Despite its promise, AI in property prediction faces significant challenges. Key issues include the quality and diversity of training data, model interpretability, and generalizability across different polymer classes. Most AI models are highly dependent on the quality of data and may fail when exposed to unfamiliar polymer systems [16]. Furthermore, the "black-box" nature of deep learning models often makes it difficult to understand the reasoning behind predictions, posing hurdles in regulatory approval and scientific acceptance [17].

5 Applications of AI in Polymer Synthesis and Processing

5.1 AI in Developing New Synthetic Routes and Processes for Polymerization

Artificial Intelligence (AI) is transforming the way chemists and materials scientists design synthetic routes for polymerization. In traditional polymer chemistry, identifying a viable synthetic route requires extensive trial-and-error experiments, expert intuition, and access to vast literature. AI simplifies this process by employing machine learning (ML), deep learning (DL), and natural language processing (NLP) to scan large reaction databases and extract useful chemical patterns [18]. Tools like the Molecular Transformer and the IBM RXN platform use transformer-based deep learning architectures to predict reaction outcomes and suggest retrosynthetic pathways, helping chemists to explore new polymer backbones and functional architectures [19]. These platforms can handle thousands of candidate reactions per second, enabling rapid innovation in developing specialty polymers for applications such as drug delivery systems, biodegradable packaging, and smart textiles. Moreover, reinforcement learning models allow AI systems to iteratively improve synthesis plans by simulating polymerization reactions and selecting conditions that maximize yield, reduce toxicity, or improve material functionality [20]. AI-guided retrosynthetic planners have even been applied to the synthesis of stimuli-responsive and bio-based polymers, where complexity makes traditional synthesis planning inefficient [21]. Autonomous laboratories equipped with AI and robotic platforms are now able to execute polymer synthesis experiments based on AI-predicted pathways, leading to accelerated discovery cycles [22].

5.2 Machine Learning in Predicting the Outcomes of Polymer Processing

Polymer processing, such as extrusion, mouldings, spinning, and 3D printing, involves complex thermomechanical dynamics that influence final material properties like tensile strength, elasticity, crystallinity, and surface finish. ML techniques are increasingly being used to model and predict these outcomes based on processing parameters. For example, in injection mouldings, machine learning models can predict defects like warping, short shots, and sink marks by analysing data such as mould temperature, injection speed, pressure, and cooling time [22].

These predictions help manufacturers proactively adjust parameters to reduce waste and improve yield. Random forest, support vector machines, and deep neural networks are among the popular ML algorithms used in processing simulations. In the context of additive manufacturing (3D printing), ML assists in optimizing printing parameters such as layer height, infill density, and printing temperature to enhance structural integrity and dimensional accuracy of printed polymer parts [23]. Inverse design methods, enabled by deep generative models, can now reverse-engineer a desired polymer property (like elasticity or thermal stability) and suggest a suitable set of printing or processing conditions to achieve it. Moreover, ML models trained on high-throughput processing data have been applied to predict melt flow index, glass transition temperature, and orientation of polymer chains during processing. This helps in tailoring polymers for niche applications, such as conductive films for flexible electronics or fibre-reinforced composites for automotive components [24].

5.3 AI-Based Real-Time Monitoring and Control During Polymer Production

The integration of AI with real-time monitoring technologies has significantly improved the efficiency, safety, and sustainability of polymer production lines. Traditionally, polymer production relied on manual monitoring and periodic sampling. However, AI now enables predictive monitoring by analysing data from sensors in real time and adjusting processing parameters dynamically. Neural networks and deep learning models are deployed to track critical variables like temperature, viscosity, pressure, and reaction rate in polymerization reactors. These models can identify subtle anomalies or deviations and trigger alerts or corrective actions before any physical fault occurs, thereby reducing downtime and improving product consistency [25].

For instance, in batch or semi-batch reactors, AI systems can optimize monomer feed rate and initiator concentration to maintain consistent polymer properties across production batches [26]. Digital twin technology AI-powered virtual replicas of physical production systems is increasingly used in polymer plants to simulate and optimize operations without interrupting actual production. These systems are capable of "learning" from historical data and improving over time, creating a continuous feedback loop that enhances process efficiency and reduces energy consumption [27]. Additionally, explainable AI (XAI) frameworks have started gaining prominence in process control, offering transparency into decision-making processes, which is crucial for quality assurance and regulatory compliance in industries like food packaging and pharmaceuticals. Some AI platforms now feature closed-loop control capabilities, where the system can autonomously run, monitor, and refine the synthesis and processing of polymers in real time. These platforms combine spectroscopy, thermal sensors, and ML algorithms to maintain optimal conditions and improve reproducibility, pushing the boundaries of what autonomous materials production can achieve [28].

5.4 AI in Specialized Polymer Applications

As polymer science evolves, specialized applications such as smart polymers, biodegradable plastics, conductive polymers, and biomedical polymers demand a more nuanced design approach. AI technologies are proving essential in accelerating development and improving performance in these niche areas.

5.5 Smart and Responsive Polymers

Smart polymers materials that respond to environmental stimuli like temperature, pH, or light have applications in sensors, actuators, and drug delivery. AI models, particularly deep learning and genetic algorithms, are employed to design responsive polymers with tailored stimulus-responsiveness. By training models on datasets comprising structural motifs and their corresponding responses, researchers can predict new configurations that exhibit specific shape-memory or self-healing properties [29]. AI further aids in simulating real-time behaviour of such polymers under varying stimuli conditions, reducing the reliance on experimental testing [30].

5.6 AI in Biodegradable and Green Polymers

Environmental sustainability is a key concern in modern polymer research. AI enables the identification and optimization of biodegradable polymer structures by predicting degradation pathways, microbial compatibility, and end-of-life behaviour. Machine learning models help screen biopolymers for composability and lifecycle performance, making it easier to align materials with circular economy goals [31]. Generative AI can also recommend bio-based monomers derived from agricultural waste or renewable sources, leading to eco-friendly innovations [32].

5.7 Conductive and Electronic Polymers

Conductive polymers like polyaniline or PEDOT: PSS are central to wearable electronics, organic photovoltaics, and flexible sensors. Their performance is highly dependent on molecular alignment, dopant levels, and synthesis conditions. AI techniques especially convolutional neural networks and unsupervised clustering are used to correlate structural properties with electronic performance metrics such as charge mobility and conductivity [33]. AI also assists in simulating charge transport behaviour in complex polymeric systems, allowing faster optimization for specific electrical applications [34].

5.8 *Biomedical Polymers and AI-Assisted Drug Delivery*

Biomedical applications of polymers such as scaffolds for tissue engineering, implant coatings, or drug delivery vehicles benefit greatly from AI-enhanced design. AI models support the prediction of biocompatibility, toxicity, and in-vivo degradation behaviour of candidate polymers. Tools using natural language processing can even mine biomedical literature to suggest new polymer drug pairings for targeted delivery [35]. Furthermore, reinforcement learning aids in selecting polymer matrices that can modulate drug release kinetics in response to physiological signals [36].

5.9 *AI for Polymer Blends and Composites*

Polymer composites and blends are crucial in aerospace, automotive, and high-strength structural materials. AI is applied in phase prediction, miscibility studies, and predicting mechanical and thermal behaviour of multi-component systems. Bayesian optimization and data-driven simulation models reduce the experimentation required to find optimal combinations of polymers, fillers, and compatibilizers [37]. AI-driven design platforms can also predict interfacial adhesion properties and fibre dispersion in polymer nanocomposites, improving durability and strength [38].

5.10 *Advanced Textile and Coating Polymers*

AI plays a role in designing polymers for functional textiles such as antimicrobial, water-repellent, or UV-blocking coatings. ML models are used to predict the interaction of textile polymers with various chemical agents and environmental stressors [39]. In advanced coatings, AI helps optimize polymer formulations for parameters like scratch resistance, gloss, and drying time, which are critical for both industrial and consumer applications [40].

6 Challenges in Integrating AI with Polymer Science

6.1 *Limited High-Quality Data*

A major obstacle in merging AI with polymer science is the lack of large, structured, and standardized datasets. Most polymer data remains scattered across publications in unstructured formats, often inaccessible or not digitized properly. Furthermore, experimental variability and inconsistencies in measurement methods introduce noise, reducing data reliability for training accurate machine learning (ML) models. This data scarcity and fragmentation significantly limit development of robust, generalizable AI models [5].

6.2 Complexity of Polymer Structures

Unlike small molecules, polymers exhibit structural complexity due to factors like polydispersity, branching, cross-linking, and conformational variability. Representing such molecular features computationally remains a challenge for conventional AI models that rely on fixed descriptors or fingerprints. New AI-ready polymer representations (e.g., graph neural networks or sequence-based models) are still evolving & require extensive domain knowledge to implement effectively [42].

6.3 Lack of Interpretability

One of the most persistent concerns in applying deep learning to polymer science is the black-box nature of AI models. While these models may generate accurate predictions, they often lack explainability, making it difficult for chemists and materials scientists to trust or validate them experimentally. This opacity hinders their adoption in safety–critical applications and reduces their utility in guiding rational material design [42].

6.4 Generalizability and Transferability

AI models trained on narrow, domain-specific data often perform poorly when applied to new polymers, formulations, or process conditions. This lack of transferability limits the scalability of AI solutions in industrial R&D settings, where novel chemistries are frequently explored. Robustness and cross-domain generalization remain underdeveloped in current models [44].

6.5 Integration into Experimental Workflows

Despite algorithmic advancements, most AI tools are not seamlessly integrated into lab or production workflows. Bridging the gap between computational prediction and real-time experimentation requires compatible data formats, standardized instrumentation interfaces, and user-friendly platforms resources that are still lacking in many polymer labs. Without such integration, AI remains underutilized in routine practice [44].

6.6 Evolving Infrastructure and Tools

Current AI applications in polymer science often suffer from inadequate infrastructure such as outdated lab software, disconnected databases, and poor automation. High-throughput experimentation (HTE), cloud-based platforms, and robotic labs are gradually being introduced, but remain inaccessible for many research groups due to cost and technical complexity [45].

6.7 Skills Gap in the Polymer Community

There exists a knowledge and skills gap between polymer scientists and AI practitioners. Many chemists lack training in data science or machine learning, while AI experts may not fully understand the physical chemistry and processing nuances of polymers. This gap hinders effective collaboration and slows the pace of innovation [46].

6.8 Data Privacy and Proprietary Restrictions

Industrial polymer data is often proprietary or confidential, limiting data sharing and collaboration. Companies are reluctant to share internal formulations, processing parameters, and product performance metrics. As a result, open-source datasets are rare, and model generalization across organizations becomes nearly impossible without privacy-preserving AI methods like federated learning [47].

6.9 Validation and Reproducibility Issues

Many published AI models are not easily reproducible due to lack of open code, insufficient data availability, or undocumented pre-processing steps. This poses a significant challenge for academic validation and industry adoption, especially in high-stakes sectors like biomedical or aerospace polymers. Standardized benchmarks and reproducibility protocols are still evolving [48].

6.10 Ethical and Regulatory Barriers

Finally, ethical concerns around bias in training data, lack of transparency, and regulatory uncertainties pose barriers to wide AI adoption in regulated sectors.

Ensuring fairness, explainability, and compliance with data governance regulations (e.g., GDPR, REACH) remains a priority, especially as AI-driven materials enter consumer markets [49, 50].

7 Future Prospects and Opportunities

The future of polymer science is poised for a transformative shift through the integration of artificial intelligence (AI), opening up vast opportunities across research, development, and industrial implementation. AI-powered autonomous laboratories are emerging as game-changers, where robotic platforms driven by machine learning can independently conduct polymer synthesis, testing, and optimization with minimal human intervention. This rapid iteration accelerates material discovery while reducing cost and time.

Predictive modeling and big data analytics will continue to enhance polymer informatics, enabling the design of novel polymers with tailored properties such as strength, elasticity, thermal resistance, or biodegradability before physical synthesis even begins. In specialized applications, such as personalized medicine and smart electronics, AI will facilitate the development of custom polymeric materials suited for specific patient profiles or environmental conditions. Industrial manufacturing will benefit from smart, AI-integrated production lines capable of real-time monitoring, predictive maintenance, and process optimization, fostering higher efficiency and sustainability. As these innovations evolve, educational and professional training in polymer science will increasingly incorporate interdisciplinary knowledge, blending traditional chemistry with computational tools and data science. To support these advancements, global standards for data sharing, algorithm transparency, and ethical AI deployment will be necessary to ensure safe, reliable, and equitable integration of AI technologies in the polymer sector.

8 Conclusion

The convergence of artificial intelligence and polymer science represents a powerful paradigm shift that is reshaping the landscape of material discovery, design, and manufacturing. By harnessing the capabilities of machine learning, neural networks, and data-driven modeling, researchers can now predict polymer properties, optimize synthesis pathways, and streamline processing with unprecedented speed and accuracy. These advancements not only accelerate innovation but also promote sustainability by minimizing experimental waste and energy consumption. As AI tools continue to evolve, their integration into polymer science will deepen, enabling the creation of next-generation materials for healthcare, electronics, environmental solutions, and beyond. However, realizing this potential requires collaborative efforts across disciplines, standardized data infrastructure, and ethical deployment of AI technologies. Ultimately, the synergy between AI and polymer science is set to revolutionize the field, paving the way for smarter, more efficient, and sustainable materials of the future.

References

1. Odian, G.: Principles of Polymerization. 4th ed. Wiley-Interscience (2004)
2. Flory, P.J.: Principles of Polymer Chemistry. Cornell University Press (1953)
3. Jordan, M.I., Mitchell, T.M.: Machine learning: trends, perspectives, and prospects. Science **349**(6245), 255–260 (2015)
4. Butler, K.T., Davies, D.W., Cartwright, H., Isayev, O., Walsh, A.: Machine learning for molecular and materials science. Nature **559**(7715), 547–555 (2018)
5. Ramprasad, R., Batra, R., Pilania, G., Mannodi-Kanakkithodi, A., Kim, C.: Machine learning in materials informatics: Recent applications and prospects. NPJ Comput Mater. **3**(1), 54 (2017)
6. Pilania, G., Wang, C., Jiang, X., Rajasekaran, S., Ramprasad, R.: Accelerating materials property predictions using machine learning. Sci. Rep. **3**, 2810 (2013)
7. Tshitoyan, V., Dagdelen, J., Weston, L., Dunn, A., Rong, Z., Kononova, O., et al.: Unsupervised word embeddings capture latent knowledge from materials science literature. Nature **571**(7763), 95–98 (2019)
8. Kim, E., Huang, K., Saunders, A., McCallum, A., Ceder, G., Olivetti, E.: Materials synthesis insights from scientific literature via text extraction and machine learning. Chem. Mater. **29**(21), 9436–9444 (2017)
9. Zunger, A.: Beware of plausible predictions of fantasy materials. Nature **566**(7743), 447–449 (2019)
10. Jablonka, K.M., Ong, S.P., Jain, A.: Using machine learning to accelerate materials discovery. Nat. Rev. Mater. **6**(8), 1–20 (2021)
11. Choudhary, K., DeCost, B.: Machine learning for materials design and discovery. J. Phys. D Appl. Phys. **54**(33), 333002 (2021)
12. Mannodi-Kanakkithodi, A., Pilania, G., Ramprasad, R.: Machine learning strategy for accelerated design of polymer dielectrics. Sci. Rep. **6**, 20952 (2016)
13. Isayev, O., Oses, C., Toher, C., Gossett, E., Curtarolo, S., Tropsha, A.: Universal fragment descriptors for predicting properties of inorganic crystals. Nat. Commun. **8**, 15679 (2017)
14. Lookman, T., Balachandran, P.V., Xue, D., Yuan, R.: Active learning in materials science with emphasis on adaptive sampling using uncertainties for targeted design. NPJ Comput Mater. **5**, 21 (2019)
15. Xie, T., Grossman, J.C.: Crystal graph convolutional neural networks for an accurate and interpretable prediction of material properties. Phys. Rev. Lett. **120**(14), 145301 (2018)
16. Schütt, K.T., Arbabzadah, F., Chmiela, S., Müller, K.-R., Tkatchenko, A.: Quantum-chemical insights from deep tensor neural networks. Nat. Commun. **8**, 13890 (2017)
17. Ghiringhelli, L.M., Vybiral, J., Levchenko, S.V., Draxl, C., Scheffler, M.: Big data of materials science: critical role of the descriptor. Phys. Rev. Lett. **114**(10), 105503 (2015)
18. Schwaller, P., Laino, T., Gaudin, T., Bolgar, P., Hunter, C.A., Bekas, C., Lee, A.A.: Molecular transformer: a model for uncertainty-calibrated chemical reaction prediction. ACS Cent. Sci. **5**(9), 1572–1583 (2019)
19. Schwaller, P., Petraglia, R., Zullo, V., Nair, V.H., Haeuselmann, R.A., Pisoni, R., et al.: Predicting retrosynthetic pathways using transformer-based models and a hyper-graph exploration strategy. Chem. Sci. **11**(12), 3316–3325 (2020)
20. Granda, J.M., Donina, L., Dragone, V., Long, D.L., Cronin, L.: Controlling an organic synthesis robot with machine learning to search for new reactivity. Nature **559**(7714), 377–381 (2018)
21. Kwon, O.H., Kim, H.S.: Machine learning-based prediction of melt flow behavior of polymers in injection molding. Polymers **12**(4), 890 (2020)

22. Tao, Y., Kou, H., Liu, Y., Jiang, X., Li, Z.: Application of machine learning in intelligent injection molding: a review. Adv. Ind. Eng. Polym. Res. **4**(2), 107–119 (2021)
23. Tlegenov, Y., Hong, G.S., Liang, S.Y.: Machine learning-based framework for predicting 3D printing process parameters and mechanical properties. J. Manuf. Sci. Eng. **141**(12), 121005 (2019)
24. Gao, W., Zhang, Y., Ramanujan, D., Ramani, K., Chen, Y., Williams, C.B., et al.: The status, challenges, and future of additive manufacturing in engineering. CAD Comput. Aided Des. **69**, 65–89 (2015)
25. Imtiaz, S., Abbas, N., Ullah, S., Khan, A., Mehmood, T.: AI-driven real-time process optimization for batch polymerization reactors. Comput. Chem. Eng. **162**, 107761 (2022)
26. Yang, Y., Luan, J., Li, B., Zhang, H.: Intelligent design of bio-based polymers using deep generative models. J. Mater. Chem. A. **10**(19), 10230–10242 (2022)
27. Ren, F., Ward, L., Williams, T., Laws, K.J., Wolverton, C., Hattrick-Simpers, J., et al.: Accelerated discovery of metallic glasses through iteration of machine learning and high-throughput experiments. Sci Adv. **4**(4), eaaq1566 (2018)
28. Ma, R., Wu, Y., Chen, Y., Zhang, C., Yan, D.: Real-time monitoring of polymerization kinetics with neural network-based sensors. Chem. Eng. J. **414**, 128850 (2021)
29. Zhang, Y., Wu, T., Wang, Z., et al.: Machine learning design of smart polymers with programmable thermal response. Adv. Mater. **34**(12), 2108795 (2022)
30. Zhang, J., Wang, B., Zhou, Z., et al.: AI-powered simulation of smart polymer behavior under external stimuli. Polym. Chem. **12**(24), 3567–3578 (2021)
31. Choi, J., Park, Y.T., Kim, S.W., et al.: AI-assisted development of biodegradable polymers for sustainable applications. Green Chem. **24**(9), 3394–3406 (2022)
32. Liu, Y., Chen, H., Wang, C., et al.: Generative design of bio-based polymers from agricultural waste using deep learning. J. Clean. Prod. **389**, 136171 (2023)
33. Lee, J.H., Kim, D.H., Shin, H.S., et al.: Machine learning analysis of charge transport in conductive polymers. Adv. Electron Mater. **7**(3), 2000997 (2021)
34. Ahmed, S., Zhang, T., Guo, H., et al.: Artificial intelligence in the optimization of conductive polymer systems for electronics. J. Mater. Chem. C. **8**(14), 4791–4802 (2020)
35. Kumar, A., Singh, H., Tiwari, A., et al.: Deep learning-driven prediction of biocompatible polymers for controlled drug delivery. Biomed. Mater. **18**(2), 025010 (2023)
36. Li, Y., Wang, L., Zhang, X., et al.: Reinforcement learning in controlled drug delivery system design using polymer matrices. Mater. Sci. Eng. C **138**, 112535 (2022)
37. Goh, G., Hegde, G., Kumar, P., et al.: Bayesian optimization in the development of high-performance polymer blends. Polymers (Basel) **13**(9), 1510 (2021)
38. Patel, S., Bose, S., Banerjee, R., et al.: AI-enhanced modeling of nanocomposite interfacial properties in high-performance polymers. Compos. Sci. Technol. **225**, 109508 (2022)
39. Jain, R., Sharma, D., Kaur, N., et al.: Predictive modeling of textile polymer finishes using AI techniques. J. Ind. Text. **52**(5), 6792–6810 (2023)
40. Wang, T., Li, J., Song, Y., et al.: AI-guided formulation optimization of polymer-based coatings. Prog. Org. Coat. **154**, 106203 (2021)
41. Jablonka, K.M., Ong, S.P., Jain, A., Bazant, M.Z., Asta, M.: Bias-free data and machine learning in materials science. Nature **592**(7853), 583–585 (2021)
42. Kim, C., Chandrasekaran, A., Huan, T.D., Das, D., Ramprasad, R.: Polymer informatics: current status and critical next steps. Mater. Sci. Eng. R. Rep. **144**, 1–23 (2018)
43. Dey, S., Choudhary, A., Agrawal, A.: Challenges in machine learning applications to materials science. JOM **73**, 2040–2045 (2021)
44. Tehrani, A.M., O'Connor, M.B., Brgoch, J.: Machine learning directed search for ultrahard materials. J. Am. Chem. Soc. **140**(31), 9844–9853 (2018)
45. Sanchez-Lengeling, B., Aspuru-Guzik, A.: Inverse molecular design using machine learning: generative models for matter engineering. Science **361**(6400), 360–365 (2018)
46. Wang, Z., Chen, X., Wang, W., Lin, S., Lin, Y.: Towards interpretable and trustworthy AI in materials science. Patterns **4**(7), 100692 (2023)

47. Cao, B., Su, Y., He, Y., Qian, Y., Jiang, Y.: AI-assisted design in polymer processing: integration strategies and challenges. Adv. Ind. Eng. Polym. Res. **5**(1), 56–67 (2022)
48. Schweidtmann, A.M., Clayton, A.D., Holmes, N., Bradford, E., Bourne, R.A., Lapkin, A.A.: Machine learning meets continuous flow chemistry: automated optimization towards the Pareto front of multiple objectives. Chem. Eng. J. **352**, 277–282 (2018)
49. Wodo, O., Ziabari, A., Ganapathysubramanian, B.: Microstructural informatics for polymer solar cells: data-driven microstructure–property–processing relationship. Comput. Mater. Sci. **125**, 203–211 (2016)
50. Liu, Y., Zhao, T., Ju, W., Shi, S.: Materials discovery and design using machine learning. J Materiomics. **3**(3), 159–177 (2017)

The Future of Polymer Science: Artificial Intelligence and Beyond

Akshita, Devang Shukla, Rohit Verma, Tejendra K. Gupta, Deepshikha Gupta, Ravi Kant Choubey, Sunil Kumar, and Vijay Kumar

Abstract The field of polymer science, which has historically relied on empirical methods and intuition-led design, has been slow to embrace data-driven paradigms. Even though artificial intelligence (AI) is offering powerful methodologies for data analysis, property prediction, and process optimization, its integration into polymer science is still limited due to the inherent complexity of polymer systems, technical difficulties, and a lack of cross-disciplinary expertise. This viewpoint offers polymer researchers a practical starting point by describing practical AI applications, easily accessible computational tools, and fundamental implementation difficulties. It seeks to demystify AI, lower adoption obstacles, and foster discussion at the nexus of polymer chemistry and machine learning rather than acting as a specialized treatise. Early AI engagement will boost creativity and discovery while putting polymer scientists at the forefront of next-generation materials research as the field develops.

Keywords Polymer science · Artificial intelligence (AI) · Data-driven materials research · Machine learning applications

1 Introduction

Polymeric materials have advanced significantly during the last century and are now essential components of contemporary technological infrastructure. These days, they are useful in a wide range of industries, such as electronics [1, 2], biomedicine [3], healthcare [4, 5], energy [6], and construction [7]. In terms of production volume,

Akshita · D. Shukla · T. K. Gupta (✉) · D. Gupta
Department of Chemistry, Amity Institute of Applied Sciences, Amity University Uttar Pradesh, Noida, India
e-mail: tejendra.amu@gmail.com

R. Verma · R. K. Choubey
Department of Applied Physics, Amity Institute of Applied Sciences (AIAS), Amity University, Noida Campus, Noida, Uttar Pradesh, India

S. Kumar · V. Kumar
Department of Physics, Indira Gandhi University, Meerpur (Rewari), Haryana, India

B. Sharma et al. (eds.), *AI-Driven Polymer Design*, Engineering Materials,
https://doi.org/10.1007/978-981-95-8353-9_10

commodity polymers like polyethylene, polypropylene, and polystyrene have even overtaken conventional metals like steel [8]. Nevertheless, these materials' performance in demanding applications is constrained by their intrinsically low mechanical strength and thermal stability. The creation of high-performance polymers has been a key area of attention in modern materials science in order to overcome these constraints [9]. A comprehensive and iterative strategy that includes controlled synthesis, scalable processing techniques, and molecular design is required for this undertaking [10, 11]. Usually lasting more than ten years, the innovation cycle necessitates a great deal of investment and study. The rate of innovative polymer discovery is hampered by the fact that traditional materials research mostly uses experience-driven, trial-and-error approaches, which are fundamentally ineffective for navigating the huge, high-dimensional, and nonlinear chemical space [12]. Data-driven and effective research paradigms are desperately needed to get over these obstacles. Significant advancements in materials science have been sparked by the recent explosion in artificial intelligence (AI) technologies [13]. Clarifying the links between processing, structure, property, and performance (PSPP), which control material behavior and functionality, is essential to this field [14]. Traditionally, experimental methods—which are frequently time-consuming and resource-intensive—are used to study such interactions. By incorporating AI into this framework, PSPP correlations may be explored quickly and in large quantities, greatly increasing the speed, scalability, and accuracy of materials discovery. New developments have shown that AI-driven approaches can be widely used to a wide range of material fields, such as energy materials [15–17], mechanical and structural materials [18], biomimetic systems [19–21], and medical materials [22], among others [23–25]. These methods make it easier to optimize composition [26], lower the cost of experiments [27], and predict properties like Young's modulus [28], melting point [29, 30], thermal stability, and thermal conductivity [31, 32] with accuracy. Additionally, AI improves characterisation methods such as X-ray absorption spectroscopy (XAS), small-angle scattering (SAS), inelastic neutron scattering (INS), pair distribution function (PDF), and powder diffraction (PD) [33]. When used in tandem, these applications not only expedite the research process but also offer strong backing for the logical design and quick development of innovative materials. Figure 1 illustrates the data-driven and multi-paradigm development processes in material science and engineering [34].

The superior mechanical, thermal, electrical, and processability characteristics of polymers are attributed to their flexible chain architectures, compositional polydispersity, sequence randomness, hierarchical multi-scale structures, and strong dependence of properties on processing parameters [35]. In contrast to inorganic materials like metals and ceramics, polymer systems show greater structural complexity and variability at the molecular level [36].

Nevertheless, they also complicate predictive modeling and greatly increase the dimensionality of the design space, making traditional trial-and-error methods inadequate for logical and effective material design [37]. A revolutionary paradigm for the structural design, property prediction, and process optimization of polymer materials has been brought about by the development of artificial intelligence (AI), with its strong generalization and feature extraction capabilities [38]. AI enables multiscale

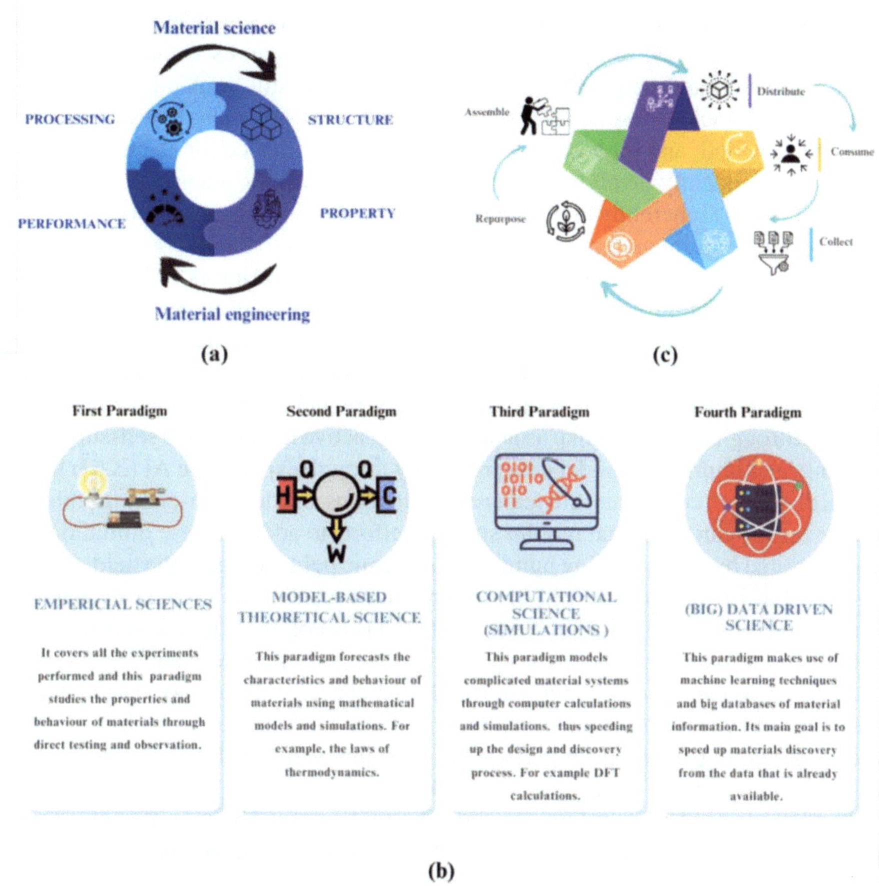

Fig. 1 Multi-paradigm development and data-driven process of material science and engineering. **a** Schematic diagram of PSPP relationship; **b** four major paradigms of material development; **c** life cycle and data of polymer materials [34]

and multivariable coordination in polymer design by making it easier to find latent, nonlinear correlations in high-dimensional PSPP (processing–structure–property–performance) spaces. The effectiveness and success rate of material development are greatly increased as a result. Utilizing AI to decipher intricate links between polymer structure and characteristics, as well as between performance measurements and processing circumstances, has become a growing area of interest in recent research. In order to accurately anticipate important characteristics like glass transition temperature, elastic modulus, and thermal stability, deep neural networks (DNNs) have been frequently used to model complex structure–property connections [39]. Simultaneously, graph neural networks (GNNs) have demonstrated efficacy in extracting chemical and topological information from polymer molecular graphs, improving the accuracy of property prediction [40]. Additionally, by adaptively modifying reaction conditions for desired performance outcomes, reinforcement learning (RL) algorithms have been used to optimize polymerization procedures autonomously. The use of artificial intelligence (AI) in polymer research is still limited by a number of significant obstacles, despite its revolutionary promise [41]. The first is the lack of consistent, varied, and high-quality datasets. Such data collection is inefficient and expensive, frequently involving lengthy lead times and labor-intensive experiments. Furthermore, the inherently multiscale and multidimensional nature of polymer structures—which can range from monomer sequences to intricate hierarchical assemblies—makes it difficult to create efficient molecular descriptors [42]. In addition, the poor interpretability of many AI models makes it difficult to derive basic scientific insights from prediction outputs and to comprehend mechanisms. Developing robust and chemically-informed descriptors that can capture the hierarchical structure–property relationships specific to polymers, reducing reliance on large, labeled datasets through techniques like transfer learning, active learning, and self-supervised learning, and improving model interpretability through explainable AI (XAI) frameworks that can offer insights into feature contributions and decision-making pathways are the three main areas that require strategic efforts to address these limitations. As demonstrated by the increasing number of academic publications in the field, the integration of artificial intelligence (AI) into polymer science has grown exponentially in recent years (Fig. 2). Machine learning (ML) is becoming a potent enabler in this field, as traditional trial-and-error approaches are becoming less and less effective in handling the enormous combinatorial design space. With previously unheard-of speed and precision, machine learning algorithms have proven to be highly effective in predicting material properties, creating innovative polymer structures, and streamlining synthesis pathways [41, 43–47].

The broad use of AI in polymer research is still hampered by a number of issues in spite of these developments [41, 43–48]. For materials scientists without computational experience, the steep learning curve of AI tools and approaches is a significant

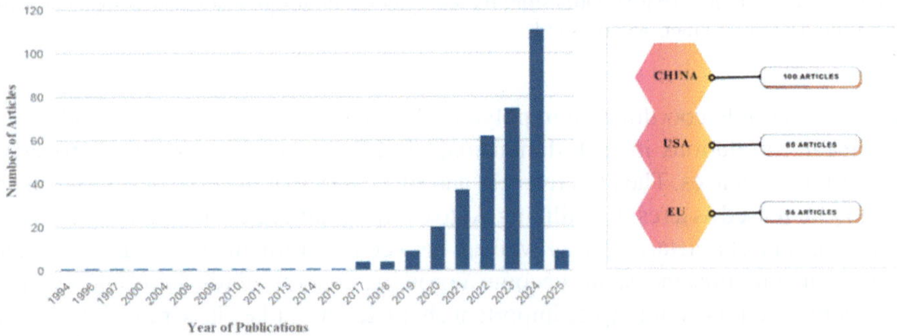

Fig. 2 Number of Publications related to AI polymer science, extracted from Web of Science using the keywords ('machine learning AND 'polymer') OR ('artificial intelligence' AND 'polymer') within the research area; material science, polymer science, and chemistry. The geographical distribution highlights the leading contributors to this emerging. EU-European Union

obstacle. Furthermore, participation from the larger polymer research community is further restricted by the lack of established workflows and easily accessible entry points. Interdisciplinary cooperation, the creation of approachable platforms, and the sharing of best practices are necessary to close this gap and promote the ethical and efficient application of AI in polymer innovation.

By giving researchers a useful starting point, this viewpoint seeks to close the gap between the fields of artificial intelligence (AI) and polymer science. This paper aims to demystify artificial intelligence (AI) and lower the obstacles to its adoption within the polymer community by highlighting important applications, fundamental machine learning (ML) techniques, and easily accessible computational tools. It is meant to serve as a foundational guide—a first step toward a deeper engagement with AI-driven materials research—rather than providing a full overview of the topic. There are a number of recent evaluations that provide insightful information for individuals looking for additional technical detail.

A thorough review of machine learning integration in self-driving labs is given by Aspuru-Guzik and colleagues [49], who emphasize the contributions of Bayesian optimization, closed-loop automation, and decision-making algorithms to the acceleration of molecular and materials discovery. Simultaneously, Stenzel et al. [50] offer a polymer-specific viewpoint, tackling important problems like data curation, the creation of machine-readable molecular descriptors, and the real-world application of ML techniques for synthetic route planning, property prediction, and new biomedical applications.

2 AI in Polymer Science as a New Scientific Paradigm

The function of artificial intelligence (AI) in polymer science is sometimes misinterpreted, especially by researchers who are not familiar with its underlying theories. AI is commonly confused with traditional simulation methods or industrial automation systems in domains that have historically relied on deterministic modeling and rule-based computation. The essential difference between expressly programmed digital tools and data-driven intelligence may be lost due to this conceptual uncertainty [51, 52]. Artificial intelligence (AI) systems discover patterns from data without explicit programming, in contrast to traditional computational models like molecular dynamics or finite element simulations, which are based on first-principles or predetermined physical laws. Similar to this, industrial automation systems lack the adaptive learning and generalization skills that characterize true artificial intelligence, even if they are able to carry out intricate control tasks. Misclassifying all computational techniques as artificial intelligence (AI) runs the risk of weakening the meaning of the term and creating irrational expectations about model performance, interpretability, and applicability. For AI to be successfully incorporated into polymer research and to promote knowledgeable interdisciplinary collaboration, it is imperative to clarify its distinctive features, such as statistical learning, pattern recognition, and model adaptation. Because of their ability to perform intricate tasks, deterministic computational tools are frequently confused with artificial intelligence (AI). Even while it can execute complex mathematical operations, a scientific calculator, for example, only uses preset algorithms and generates deterministic outputs, meaning that identical inputs always result in identical outputs. It is incapable of learning, adapting, or becoming better with time.

An AI-powered system, such an adaptive mathematics assistant, on the other hand, demonstrates essentially distinct capabilities. These systems have the ability to decipher handwritten equations, recommend contextually appropriate substitute solutions, and improve their results in response to previous user input. AI differs from traditional rule-based computation because of this learning-driven adaptability, which is a feature of machine learning algorithms. It highlights the special usefulness of AI in dynamic, data-rich situations by enabling ongoing development and personalization [51, 52]. The implementation of artificial intelligence (AI) in industrial settings is not implied by the simple existence of sensors or automated equipment. Conventional automation and control systems are incapable of learning or adapting; instead, they function according to predetermined, set instructions. True AI systems, on the other hand, constantly examine data, identify intricate patterns, and react quickly to shifting circumstances. AI goes beyond simple automation in polymer research by facilitating data-driven materials discovery workflows, predictive modeling, and self-optimizing synthesis procedures.

AI is a fundamentally different paradigm from traditional computational techniques like density functional theory (DFT) [53] and molecular dynamics (MD) simulations [54], which use first-principles equations to model physical phenomena like phase behavior, chain conformations, or mechanical responses. Even without full mechanistic insight, it uses big datasets to find statistical connections and produce precise predictions. AI is positioned as a potent supplement to conventional simulation methods thanks to its data-centric approach, which speeds up hypothesis creation and optimization across intricate, multidimensional polymer systems. Many artificial intelligence (AI) models function as "black boxes," producing results based on intricate statistical correlations rather than clearly stated physical rules, notwithstanding their capacity for prediction [55]. Researchers that place a higher priority on mechanistic understanding are frequently skeptical of this opacity, which presents a serious obstacle to scientific interpretability. Acceptance may be hampered by the inability to trace AI actions back to their foundational ideas, especially in fields where explainability is crucial.

AI has, nevertheless, shown great promise in enhancing conventional modeling techniques. In contrast to molecular dynamics (MD) simulations, Bhattacharya and Patra [56] effectively used AI to forecast polymer phase transitions, such as the coil-to-globule transformation, with high accuracy and a substantially lower processing cost. AI facilitates quick screening and pattern detection in high-dimensional regions, which improves the investigation of complicated polymeric systems rather than replacing physics-based models [57]. Artificial intelligence is not a panacea, though. The quality, diversity, and representativeness of the training data are intrinsically linked to its performance. The practical dependability and resilience of AI models in real-world polymer research contexts can be limited by inaccurate or non-generalizable predictions caused by inadequate or biased datasets. Therefore, meticulous dataset validation and curation are still necessary to guarantee reliable AI applications.

3 AI Utilization in Polymer Materials

Polymer materials research is changing as a result of artificial intelligence (AI), which makes accurate prediction and optimization possible. Recent developments in three main areas are described in this section: the development of AI methodology, the use of AI to investigate structure-property correlations, and its function in processing-performance improvement.

3.1 Fundamentals and Uses of Machine Learning in Materials Science

Inspired by the architecture of human neural networks (NNs), machine learning (ML) makes it possible to find and learn intricate patterns in data, creating nonlinear mappings between input variables (like chemical descriptors) and output targets (like material properties) [58]. When properly taught, machine learning models are able to approximate "effective theories" that represent quantum mechanical principles and, in certain situations, capture underlying physicochemical relationships [59, 60]. In materials science, machine learning (ML) has been increasingly popular in recent years, especially in the areas of property prediction and the logical design of new materials. The development of chemically meaningful and information-rich descriptors and the availability of high-quality, diversified databases are two essential components for the success of machine learning in this setting. The accuracy, generalizability, and interpretability of prediction models in polymer research are all influenced by these factors taken together. Extensive datasets resulting from high-throughput simulations and experimental observations are available in databases like the Materials Project [61], AFLOW (Automatic-FLOW for Materials Discovery) [62], the Open Quantum Materials Database (OQMD) [63], and PolyInfo [64]. These sources provide as the fundamental framework for materials science machine learning (ML) model training, validation, and benchmarking. In parallel, descriptors are essential because they translate intricate structural and compositional data, whether at the atomic, molecule, or macromolecular level, into numerical representations that machine learning algorithms can understand. Uniqueness, discrimination, computational accessibility, and physical meaning are all necessary for effective descriptors. Molecular fingerprints [65], structural similarity indices, topological metrics, molecular orbital energies, atomic electronegativity, molecular weight, and thermophysical characteristics like glass transition temperature [48] are examples of frequently used descriptors. Accurate structure-property prediction and the capture of the inherent properties of polymer systems depend on these traits. Support vector machines (SVMs) [66, 67], random forests (RFs), and kernel ridge regression (KRR) are examples of traditional machine learning (ML) techniques that have shown significant effectiveness in modeling low-dimensional and structured datasets. However, when applied to high-dimensional or unstructured data sources like time series, pictures, or spoken language, these techniques frequently show inadequate scalability and generalization. Deep learning (DL), a branch of machine learning (ML) distinguished by its improved representational capacity and capacity to model intricate, nonlinear patterns, has emerged as a result of this constraint. The effective training of deep neural networks (DNNs) using graphical processing units (GPUs) for image classification tasks in 2011 by Dan Ciresan and colleagues marked a crucial turning point in the development of DL. Hinton, Bengio, and LeCun's groundbreaking work developing multilayer neural architectures that could learn hierarchical characteristics from data [68] served as the basis for this development.

Since then, DL designs including graph neural networks (GNNs), recurrent neural networks (RNNs), and convolutional neural networks (CNNs) have shown impressive performance in a variety of disciplines. A crucial benefit for materials informatics and polymer research, these models are especially good at capturing intricate connections in sizable, high-dimensional datasets since they automatically extract multilevel representations. Various categories of machine learning algorithms are mentioned in Table 1.

Natural language processing (NLP), computer vision, and speech recognition have all seen tremendous breakthroughs thanks to the quick development of computing hardware, the exponential growth of data availability, and ongoing improvements in training algorithms like the Adam optimizer, dropout regularization, and batch normalization. Deep learning (DL) has since been included into materials science research as a result of these advancements. This has led to the emergence of a vast array of machine learning (ML) techniques (as listed in Table 1), which give researchers strong tools to speed up materials design and discovery [70–76]. In materials research, machine learning (ML) greatly cuts down on the time and expense of experimental synthesis and characterization, allowing for performance prediction, inverse design, and high-throughput screening. Because DL can learn complicated, multiscale representations from vast datasets, it has been widely used for generative modeling of novel materials, property prediction, and synthetic pathway optimization. Large-scale DL model training involves significant computing demands, but their scalability and ability to extract hierarchical features offer a significant benefit when working with high-dimensional data. Furthermore, the range of AI applications in polymer science is growing quickly, combining techniques from automated experimentation, computer vision, and natural language processing. AI is being used more and more in materials research at every stage, including design, diagnostics, data analysis, predictive maintenance, and early fault detection, as we move into the Industry 4.0 era. This is opening the door for more independent, clever, and effective research workflows.

Table 1 Categorization of machine learning algorithms, including both classical and latest development algorithms [69]

Class of algorithm	Classic important algorithms	Latest development algorithm
Supervised learning	Gaussian process regression, logistic regression, decision trees, support vector machines, random forests, multilayer perceptrons, and linear models	Neural Tangent Kernel, TabNet, LightGBM, CatBoost, and XGBoost
Unsupervised learning	Principal component analysis, hierarchical clustering, and K-means	Variational autoencoder, UMAP, Deep clustering, and Learning contrast (SimCLR, MoCo)
Semi-supervised learning	Label propagation, co-training, and self-training	Fix match, MixMatch, snorkel, consistency regularisation, and pseudo-labeling
Deep learning	Long short-term memory, generative adversarial network, recurrent neural network, and convolutional neural network	Diffusion model, graph neural networks, transformer (BERT, GPT, vision transformer)
Reinforcement learning	Q-learning, deep q-network, state-action-reward-state-action, and Monte Carlo approach	DreamerV2
Meta learning/auto ML	Bayesian optimisation, grid search, and Random search	Hyperband, meta-learning (MAML, reptile), neural architecture search

3.2 Machine Learning Interatomic Potentials

In polymer simulations, machine learning interatomic potentials (MLIPs) have greatly expanded the available spatial and temporal scales, allowing for the accurate and efficient modeling of complex systems. MLIPs are an essential link between emergent macroscopic qualities and basic polymer attributes like chemical composition, chain conformation, stereoregularity, and monomer sequence. By solving the Schrödinger equation, density functional theory (DFT) has historically offered high-fidelity insight into electronic structure; yet, its computing complexity makes it unfeasible for large-scale or long-timescale simulations. However, despite their higher efficiency, classical molecular dynamics (MD) simulations—which use empirical force fields—frequently fail to capture quantum-level interactions, particularly when bond breakage, charge redistribution, or conformational transformations are involved. This trade-off is addressed by MLIPs, which maintain MD-like computational efficiency while approximate DFT-level precision [69]. MLIPs provide a potent tool for predictive modeling and materials design in polymer science by directly learning interatomic interactions from high-level quantum mechanical information. This allows for multiscale simulations that maintain quantum precision throughout larger systems and longer timescales. The expressiveness and quality of feature descriptors used to characterize atomic and molecular environments have a significant impact on the effectiveness of machine learning interatomic potentials (MLIPs). The majority of early methods used manually created descriptors that were based on domain-specific chemical heuristics or polymer repeat units. Using these techniques, scientists manually found relationships between material characteristics and structural motifs in huge datasets [69]. The growing expansion of polymer datasets makes this manual feature engineering impossible, even though it works well for small-scale investigations.

In order to get around these restrictions, Gurnani et al. [37] automated the extraction of structural information from polymer repeat units using sophisticated deep learning (DL) architectures, such as graph neural networks (GNNs) and multi-task learning (MTL). The model can develop hierarchical, chemically-informed representations straight from molecular graphs thanks to its data-driven approach, which does away with the requirement for human interaction. In comparison to conventional techniques, their method showed a 10–100 × speedup in feature extraction while retaining similar accuracy in tasks involving property prediction. These developments demonstrate how automated, scalable feature learning is becoming increasingly important for creating high-fidelity machine learning models for polymer simulations.

With a focus on the identification of high-performance polyimide dielectric materials, Zhang et al. presented the COMFO model, a composite deep learning framework intended to improve the feature extraction procedure for polymer property prediction [69]. In order to capture complete molecular representations, the model combines three different feature extraction strategies: (i) a bidirectional encoder transformer that extracts sequence-level information from simplified molecular-input line-entry system (SMILES) sequences; (ii) an attentive fingerprint (FP) network that extracts atomic and bond-level features from molecular graphs; and (iii) conventional molecular fingerprints that capture substructural motifs. COMFO greatly increases feature extraction efficiency over conventional, single-modality methods by utilizing these complimentary feature representations. Interestingly, it speeds up the discovery of promising candidate materials while maintaining excellent predictive accuracy. The efficiency of incorporating cutting-edge deep learning algorithms for quick, precise, and scalable polymer informatics is demonstrated by this multi-representational approach.

A crucial factor in determining MLIP performance is the selection of the machine learning (ML) algorithm, in addition to the quality of the descriptor. A number of end-to-end learning frameworks have been proposed to bridge the gap between large-scale molecular dynamics and quantum-level accuracy. These frameworks avoid the computationally demanding Kohn–Sham equations of density functional theory (DFT) by directly predicting electronic properties, like charge densities, from atomic configurations. One well-known example is the model created by del Rio et al. [77], which considerably lowers computing cost while achieving near ab initio accuracy. When simulating huge and complex polymer systems that are otherwise unmanageable using traditional DFT approaches, this method works especially well. Enhancing traditional molecular dynamics (MD) simulations by adding machine-learned or auxiliary energy terms to increase the fidelity of interatomic potentials is an alternate but complementary approach. The TensorMol model, for example, was described by Parkhill and associates [78] and combines many-body expansions with neural networks to capture long-range interactions, like dispersion and Coulombic forces, that are sometimes not well captured in conventional DFT. The physical realism and prediction power of simulations are significantly enhanced by these algorithmic developments, allowing for more precise and effective investigation of polymer behavior at various scales. TensorMol improves the accuracy and transferability of molecular simulations across chemically varied polymer systems while simultaneously drastically lowering computing overhead [78]. These sophisticated models are a crucial step in balancing mesoscale and macroscale modeling with quantum-level accuracy. Nevertheless, there are still issues with their numerical stability in heterogeneous material systems and their generalizability across wide chemical spaces, despite their encouraging potential. These approaches are now mostly in the exploration and optimization stages. However, they offer a strong conceptual and computational foundation for the creation of universal, data-driven models in the future that can smoothly combine mechanical behavior and electrical structure in polymers.

3.3 Inverse Polymer Design and Structure-Property Modeling

Designing materials with adjustable performance and programmable functions requires a thorough understanding of the mechanisms that link polymer structure to macroscopic features. In this regard, property-guided inverse design and structure-property modeling have become essential polymer science techniques. Predicting local physicochemical reactions from known structural configurations has been the focus of recent attempts. In order to model the potential of mean force (PMF) between polymer-grafted nanoparticles, Gautham et al. [79] created a deep neural network (DNN) architecture that includes Behler–Parrinello symmetry functions. To capture geometric and chemical surroundings, their method makes use of radial and angular symmetry descriptors that are focused on the centroid of the nanoparticle and its grafting sites. A dual-network architecture is used, with one DNN mapping the potential of the local chemical environment and the other predicting the potential of the global interaction. A high-resolution model of interparticle forces in intricate soft-matter systems is provided by this technology, which successfully represents the anisotropic interactions resulting from polymer grafting.

The DNN-based approach successfully overcomes the drawbacks of conventional models in capturing angle-dependent interactions to improve generalizability, increasing computational efficiency and setting the stage for the creation of universal models of interatomic interactions. In a related study, Shi et al. used support vector regression (SVR) with coarse-grained molecular simulations to develop a prediction framework for calculating adhesive free energy from polymer sequences. By avoiding the necessity for complex theoretical approximations, this hybrid framework drastically lowers processing cost [69]. Additionally, the methodology facilitates the inverse design of functional polymer systems by combining genetic algorithms (GA) with the SVR model to enable high-throughput screening of polymer sequences suited to particular sticky qualities. The framework's application to more extensive interfacial material design is nevertheless intriguing, despite its limited generalizability and applicability to complicated real-world systems. In a different development, Yang et al. created the Fiber Network Deep Learning System, a method based on convolutional neural networks (CNNs) that can extract mechanical and structural data from fiber network atomic force microscopy (AFM) images [69]. This model provides a strong tool for the automated analysis and design of fibrous polymeric materials by utilizing deep learning to enable quantitative assessment of network shape and mechanical behavior.

By successfully resolving issues related to poor image segmentation, the Fiber Network Deep Learning System makes it possible to precisely extract morphological features including node distribution and fiber contour length. The technique shows its validity in silk fibroin nanofiber systems and measures mechanical indicators such as persistence length by including the worm-like chain model. This method lays the groundwork for "image-driven material design," in which mechanical and structural predictions are informed by quantitative image data. Fang et al. [80] extended this paradigm by analyzing block polymer topologies using small-angle X-ray scattering (SAXS) profiles using a physics-informed machine learning approach. In morphological classification tasks, the model obtained an out-of-sample prediction accuracy of almost 95% by identifying chemical-agnostic characteristics from scattering data. When combined, these developments make it possible to create strong, broadly applicable structure–property relationships, which greatly improves the precision of predictions for mechanical and interfacial properties. This expedites the design of experiments and optimizes the choice of materials for desired functionality. By concentrating on inverse design—the process of determining the ideal material structures from goal properties—another important research avenue goes beyond forward modeling. Extensive search tactics are computationally costly due to the combinatorial complexity of polymer design spaces, which include variations in monomer sequences and copolymer compositions. Performance-driven structural optimization is thus made possible by data-driven inverse design, which offers an effective substitute. In order to intelligently regulate polymer molecular weight distributions, Li et al. [81] demonstrated this strategy by combining reinforcement learning (RL) with atom transfer radical polymerization (ATRP) simulations. Their framework introduces an AI-driven paradigm for "inverse synthetic design" and shows how to synthesize functional materials in a scalable and effective manner. The model highlights the potential of cross-disciplinary approaches that combine machine learning and polymer chemistry, even though it is subject to some simplifications and needs more extensive experimental validation. Similar to this, Kranthiraja et al. [82] used a random forest (RF) model to screen a large molecular design space using a comparatively small dataset of 566 experimental samples for the creation of organic photovoltaic materials. This method highlights the significance of machine learning (ML) in speeding up the discovery of next-generation optoelectronic polymers by demonstrating the viability of using minimal data to conduct high-throughput, performance-oriented material screening. Recent efforts have built a strong, experimental-oriented, data-driven paradigm that is changing polymer research, despite persistent issues with descriptor selection and model interpretability. To accomplish targeted polymer sequence design, Webb et al. [83] showed a strong combination of machine learning (ML) and coarse-grained modeling. Their method kept target property variations under 10% and maintained computational efficiency while achieving prediction accuracies of over 95% across a variety of polymer classes. This system offers a scalable solution to important bottlenecks in polymer design and demonstrates how machine learning (ML) has the ability to drastically lower the computing cost of exploring intricate molecular design regions.

A groundbreaking deep learning (DL)-based inverse design system specifically designed for polymer alloys was presented by Hiraide et al. [69], building on this momentum. They made it possible to directly infer ideal phase-separated microstructures from desired mechanical stress profiles by combining convolutional neural networks (CNNs) with conditional generative adversarial networks (cGANs). This architecture offers a new technique to develop materials with customized mechanical performance and marks a revolutionary advancement in structure–property inverse modeling. When taken as a whole, these data-centric frameworks highlight the move toward intelligent, high-throughput material design and have the potential to speed up the creation of high-performance, next-generation polymers. Achieving a high coefficient of determination ($R2 > 0.9$), the integrated framework created by Hiraide et al. demonstrates exceptional effectiveness in both mechanical property prediction and structural production, as measured by precise structure factor matching. The paradigm change from empirical trial-and-error methods to data-driven, predictive materials engineering is advanced by this dual-capability model, which provides a reliable and scalable path for the logical design of high-performance materials. In order to support this shift, Zhu et al. [84] presented a data-driven soft sensor built on a deep belief network (DBN) that is especially intended for industrial polymer manufacturing settings with unbalanced datasets and nonlinear behaviors. The deployment of deep learning (DL) models in actual chemical process systems reached a major milestone when the DBN model showed exceptional real-time performance in quality monitoring and control.

A data-integrated inverse design architecture was concurrently proposed by Sattari et al. [85], which combines experimental observations with high-fidelity simulation data in a synergistic manner. This closed-loop system speeds up the development and application of functional polymeric materials by enabling effective optimization throughout the whole polymer development process, from molecular design to manufacturing (Fig. 3). When taken as a whole, these developments highlight how artificial intelligence has the power to revolutionize polymer science research and industrial processes.

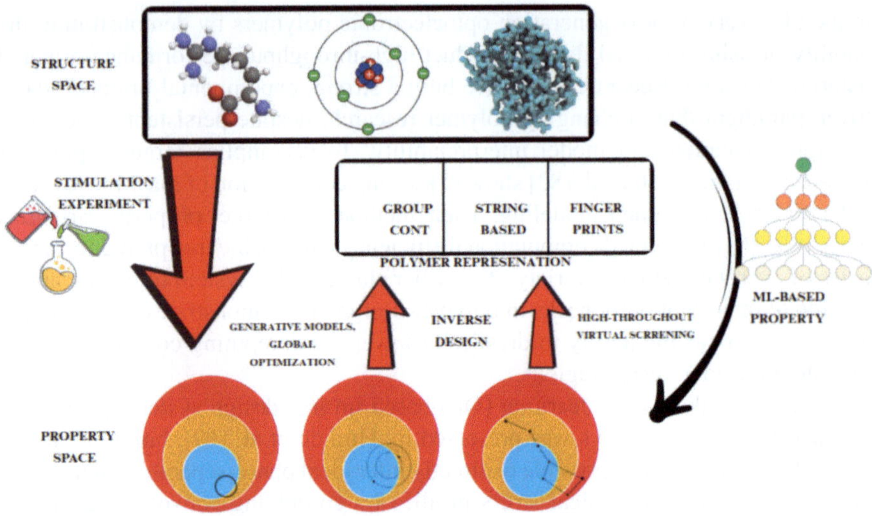

Fig. 3 Schematic diagrams of forward and reverse material designs

3.4 Properties Prediction Model

Machine learning (ML)-based intrinsic property estimate is a significant parallel research path to local response prediction in the structure–property modeling framework for polymeric materials. Macroscopic, system-level features known as intrinsic properties—such as glass transition temperature (Tg), electronic bandgap, Young's modulus, dielectric constant, melt flow index, and gas permeability—have a significant impact on the performance, suitability, and dependability of materials in a variety of engineering applications [86]. Establishing precise, generalizable mappings from molecular-level representations (such as SMILES strings, molecular fingerprints, and graph-based descriptors) to scalar physical property outputs is the main goal of intrinsic property prediction. By avoiding computationally costly quantum mechanical simulations and time-consuming experimental characterisation, these machine learning methods hope to enable high-throughput virtual screening and performance-guided material selection. Such predictive frameworks greatly speed up the search and optimization of polymeric materials for desired functions by fusing domain-specific feature engineering with cutting-edge learning algorithms. The glass transition temperature (Tg) of polyhydroxyalkanoate (PHA) polymers was predicted using Random Forest (RF) and Kernel Ridge Regression (KRR) models in a groundbreaking study by Pilania et al., creating a strong structure–property mapping with good prediction accuracy [87]. Although this method had several drawbacks, especially when it came to taking configurational effects into consideration and expanding its application to structurally diverse polymer families, it much exceeded conventional empirical methods in terms of precision and generalizability. Transfer learning techniques were incorporated into graph neural networks (GNNs) by Shi et al. in order to increase model adaptability and expand prediction capability across chemical spaces. In contrast to time-dependent density functional theory (TD-DFT) computations, they were able to obtain excited-state energy estimates with errors below 200 meV by pretraining the model on short conjugated oligomers and fine-tuning it on small datasets of longer oligomers [88]. This method improved accuracy and transferability in polymer informatics by successfully resolving the data scarcity problem that arises when simulating optoelectronic properties of long-chain conjugated systems.

In polymer informatics, multitask learning (MTL) has become a potent technique for improving modeling efficiency and taking advantage of inter-property correlations. A multitask deep neural network (DNN) framework was created by Kuenneth et al. to concurrently predict 13 important polymer properties, including mechanical, thermal, and gas permeability traits [69]. The model showed strong prediction accuracy across a variety of property domains and increased training efficiency by utilizing shared representations across tasks. If the intrinsic difficulty of balancing task-specific losses is not adequately addressed, it might reduce the accuracy of individual predictions, particularly when task relevance or data quality varies. This emphasizes how sophisticated multi-objective optimization methods are required to guarantee fair model performance on all tasks. Integrating transfer learning with multi-origin data is a viable approach to mitigate this problem. Phan et al. presented a hierarchical transfer learning system in which sparse experimental data of bulk polymer characteristics was used to refine language models that had previously been trained on quantum chemistry datasets of tiny molecules [89]. By first embedding basic electronic properties (like dipole moments and polarizabilities) and then using domain-specific calibration to adjust to macroscopic behaviors (like dielectric strength and tensile modulus), this method takes use of cross-scale knowledge transfer. The framework creates a scalable and generalizable modeling paradigm that goes beyond conventional simulation-bound approaches by bridging quantum-level descriptors and continuum-scale behaviors in addition to avoiding data scarcity.

Himanshu et al. successfully addressed important limitations in deep learning models for polymer property prediction by introducing a methodical approach for optimizing neural network (NN) architectures through a layer-by-layer expansion strategy combined with a latent space–based data selection technique [69]. This dual strategy improves the efficiency and interpretability of the model, and its modular architecture enables adaptation to a variety of material and polymeric systems. Additionally, a scalable and transferable approach for improving feature extraction and predictive modeling across material domains is provided by the combination of architecture optimization and well-informed data curation. Image-based features for property inference have been investigated concurrently in an attempt to move beyond structure-based descriptors. Notably, Liang et al. achieved a high classification accuracy of 94% by using convolutional neural networks (CNNs) in combination with transfer learning to enable automated miscibility prediction in polymer blends [69]. The model provides a quantitative metric for miscibility assessment in addition to making it easier to identify miscibility from scanning electron microscopy (SEM) pictures. The approach's scalability is encouraging, and better data quality should greatly increase its resilience and application in microstructure-driven property analysis, despite the fact that its performance is now limited by the resolution and consistency of input images. The creation of sophisticated data-driven approaches for polymer informatics has been accelerated by the Materials Genome Initiative (MGI). Taking advantage of this momentum, Kim et al. presented a hierarchical fingerprinting technique that can encode polymer properties at several scales, ranging from mesoscopic morphologies to atomic-level configurations [56, 69]. This multi-scale representation made it possible to forecast a variety of polymer properties quickly and accurately when combined with machine learning methods. The created models have been included into the Polymer Genome platform, a freely available resource that makes it easier to perform inverse design and high-throughput virtual screening of polymeric materials. In order to increase the platform's usefulness across application domains, future efforts will focus on broadening the range of targeted attributes as well as the chemical design space. In parallel progress, Upadhya et al. used evidential neural networks to rationally construct single-chain polymer nanoparticles (SCNPs) by analyzing light scattering and small-angle X-ray scattering (SAXS) data [90]. 30 new compact SCNPs were found using their model-guided synthesis and characterization pathway; 58% of these had a Porod exponent ≥ 3.5, which is suggestive of compact, globular structures. In order to speed up the identification of precisely tailored polymer nanostructures, this study demonstrates how ML models may identify experimentally feasible design choices with desired structural properties.

3.5 Properties Prediction Model

Rapid identification of high-performance materials in the field of polymer material research and development is significantly hampered by major bottlenecks such the high cost of experimental resources, lengthy synthesis timelines, and the combinatorial vastness of design factors. Artificial intelligence (AI) is becoming an active participant in experimental workflows, rather than just a predictive auxiliary tool [56, 69]. AI techniques are now included into all aspects of the experimental workflow, including autonomous synthesis decision-making, multi-variable parameter optimization, and experimental design. This is a paradigm change away from traditional "model-assisted" testing and toward a more sophisticated "system-led" approach. Modern efforts can be roughly divided into two classes based on the degree of integration and the particular optimization objectives: (i) parameter optimization within static experimental protocols and (ii) system-level closed-loop experimentation, in which artificial intelligence (AI) dynamically regulates and improves the entire experimental process in real-time to speed up material discovery.

3.6 Parameter Optimization of the Process

Statistical process control and design of experiments (DOE) were the main methods used for process optimization in polymer materials prior to the development of AI-driven approaches. Long experimental cycles, labor-intensive data management, high operating costs, poor generalizability, and an inability to represent complicated nonlinear interactions among multivariate parameters are intrinsic limitations of these old methods, notwithstanding their effectiveness in organized parameter tuning. A paradigm shift in polymer research and development has been brought about by the introduction and integration of artificial intelligence, specifically machine learning (ML) techniques. The development of predictive, data-driven models that can intelligently optimize input parameters without requiring modifications to the underlying experimental frameworks has been made possible by algorithms like support vector machines (SVMs) and artificial neural networks (ANNs). In addition to improving experimental repeatability and throughput, these models have shown quantifiable gains in the functioning and quality of polymeric products. Intelligent polymer design now relies heavily on machine learning (ML)-based parameter optimization, which can be broadly divided into two areas: (1) process-level parameter control, which aims to improve polymer forming methods like additive manufacturing, and (2) structural-level parameter regulation, which focuses on improving material composition and macroscopic performance characteristics. Khanzadeh et al. [91] employed a self-organizing map (SOM) algorithm to study geometric error distributions in the fused filament fabrication (FFF) process of polymers in the context of additive manufacturing. The method eliminated the requirement for manual data labeling by automatically classifying and identifying geometric deviations from point cloud data from scanned printed objects. Notably, the model demonstrated exceptional efficiency in defect recognition and surface quality evaluation, achieving accurate mistake detection with just 2.4% of the entire dataset. The study developed a strong methodological framework for extending ML-assisted diagnostics to more complicated and fine-featured polymer components, despite its concentration on simplified test geometries, which limited its immediate generalizability to complex or industrially important structures. Roy et al. [92] created a neural network (NN) model that showed remarkable computational efficiency and prediction accuracy in the field of thermal field modeling for polymer processing. The NN generated a single temperature profile in 0.036 s, with a forecast error of less than 5%, compared to traditional finite element method (FEM) simulations. In fixed-structure fused filament fabrication (FFF) settings, the model demonstrated excellent stability and adaptability, providing crucial assistance for process optimization, real-time monitoring, and closed-loop control during additive manufacturing.

Machine learning (ML) has been widely used into several additive manufacturing processes, such as material extrusion, photopolymerization, powder bed fusion, binder jetting, and material jetting, in addition to thermal modeling. For instance, Zhu et al. [69] used support vector machines (SVM) in conjunction with principal component analysis (PCA) to reduce printing flaws in material extrusion procedures. To detect anomalies in real time during photopolymerization, Edlim et al. [69] created a convolutional neural network (CNN) architecture based on the YOLO (You Only Look Once) framework. Similar to this, Lee et al. [93] accurately identified sub-micron structural inconsistencies in two-photon lithography by combining CNN-LSTM and 3D-CNN models for defect detection. Together, these developments highlight how ML is revolutionizing dependability, accuracy, and efficiency across a range of polymer manufacturing methods. Additionally, in order to facilitate early-stage quality assessment in additive manufacturing processes, Sassman et al. [56, 69] used support vector machines (SVM) to build a pre-screening framework that made use of powder flowability and surface roughness measurements. In order to improve the accuracy of defect recognition in polymer-based fabrication processes, Satterlee et al. [69, 94] combined image enhancement techniques with a suite of convolutional neural network (CNN) designs to make it easier to detect porosity-related flaws. When taken as a whole, these investigations highlight the resilience and adaptability of deep learning (DL) models, including CNNs, recurrent neural networks (RNNs), and multilayer perceptrons (MLPs), in identifying intricate, high-dimensional connections in polymer formation processes. In addition to polymer forming, parameter optimization techniques have been widely used to improve functional performance and customize material composition. In order to analyze experimental data and optimize the dispersion ratio of graphene oxide–hydrogen-bonded zinc hydroxystannate within a polypropylene (PP) matrix, Chen et al. [69] created an integrated machine learning strategy that included the Sure Independence Screening and Sparsifying Operator (SISSO) method. The predictive model that resulted greatly improved the PP composite's flame-retardant performance and showed excellent generalizability, indicating that it may be applied to a wider variety of polymer systems.

Bayesian optimization (BO) was used by Li et al. [95] to methodically improve the synthesis parameters for the manufacturing of short polymer fibers. When compared to traditional grid search techniques, their method significantly reduced the number of experimental runs while successfully navigating high-dimensional and poorly sampled process spaces. The BO framework demonstrated its effectiveness and superiority in single-objective process optimization by enabling quick convergence towards ideal synthesis conditions. However, a thorough validation of its generalizability across various polymerization systems is still pending. Similar to this, Takasuka et al. [96] applied BO to fine-tune crucial processing parameters while synthesizing styrene–methyl methacrylate (St–MMA) copolymers using a flow synthesis platform. The work confirmed the usefulness of BO in precision copolymer engineering by successfully manipulating monomer ratios. Future-focused, data-driven, intelligent design of functionally customized copolymers is expected to be made possible by the simultaneous optimization of compositional and performance-related parameters made possible by the integration of multi-objective BO.

In conclusion, advanced machine learning frameworks like Bayesian Optimization (BO), Multi-Task Learning (MTL), and hierarchical modeling have replaced more conventional unsupervised clustering and statistical regression techniques as the primary tools for parameter optimization in polymer research. In extremely parameter-sensitive processes, such as additive manufacturing, where exact control over variables is crucial for performance optimization, these modern methods demonstrate improved efficiency and flexibility [69, 95, 96]. Additionally, they have proven to be highly applicable in a variety of fields, such as the calibration of coarse-grained force fields, composite material formulation, and copolymer composition tuning, providing a scalable and data-driven basis for logical materials design across polymer systems [69, 83].

3.7 *Closed-Loop Experiments and Process Iteration*

Recent work has concentrated on the creation of autonomous closed-loop systems that facilitate iterative experimentation, intelligent decision-making, and real-time process refinement in polymer materials engineering, building on developments in parameter tuning and process optimization. By improving experimental throughput, enabling precise process control, and facilitating more sustainable workflows, the area has undergone a significant transformation due to the confluence of automation technologies and artificial intelligence (AI) [97, 98]. Integrating deep learning (DL) algorithms with high-throughput experimental platforms is essential to this shift because it enables adaptive process iteration and feedback-driven experimental cycles. By facilitating continuous optimization across the design–synthesis–evaluation loop, decreasing human intervention, and speeding up discovery, these automated frameworks are gradually changing traditional research paradigms. Sbrana et al. [69] used the N-BEATS deep learning architecture in combination with near-infrared hyperspectral imagery to categorize over 4500 plastic samples in the material identification and sorting field. The model's overall categorization accuracy was 79%, with colored plastics having a significantly greater accuracy (90%) than black plastics (67%). This data-driven approach simplified upstream supply chain processes by improving recycling efficiency and optimizing raw material acquisition.

In order to study the catalytic kinetics of homogeneous polymerization, Rizkin et al. [99] created a semi-autonomous microfluidic platform that was combined with in situ thermographic monitoring and neural networks. When used in zirconocene-catalyzed α-olefin polymerization, this technology reduced energy consumption and chemical waste while greatly increasing reaction efficiency. The study demonstrated the feasibility of combining high-throughput microfluidics with machine learning to enhance the mechanistic comprehension and sustainability of polymerization processes, notwithstanding significant restrictions in reaction scope. In polymer research, process automation has emerged as a key factor in driving high-throughput experiments. The machine learning-guided experimental platform created by Cao et al. [100] combines design of experiments (DOE) and Bayesian optimization (BO) techniques in a complementary manner. When this platform was used to optimize complex liquid formulations, it completed hundreds of experiments in 15 working days and found nine ideal formulations that met all target requirements. This method significantly improved both experimental throughput and predicted accuracy when compared to traditional empirical procedures.

Knox et al. [101] developed an automated platform for performing reversible addition–fragmentation chain transfer (RAFT) polymerizations under regulated conditions in order to further increase synthesis precision and scalability. The system minimized experimental iterations while achieving closed-loop multi-objective optimization through the use of the Thompson Sampling Efficient Multi-objective Optimization (TSEMO) method. This platform showed great flexibility in a variety of polymerization systems with intricate input–output interactions, and it has a lot of potential for investigating polymer behavior in harsh or unusual environments. In conclusion, experimental closed-loop platforms and process iteration are quickly taking center stage in the transition of polymer research from empirically led approaches to data-centric frameworks. These solutions address recurring inefficiencies in material discovery and optimization workflows by seamlessly integrating algorithmic decision-making, autonomous execution, real-time feedback acquisition, and high-throughput experimental design. These platforms greatly improve polymer materials science's capacity for innovation, scalability, and reproducibility by facilitating adaptive learning cycles and speeding up the investigation of intricate parameter spaces [102, 103].

4 Important Methods for Machine Learning

A key area of artificial intelligence is machine learning (ML), which makes it easier to create data-driven models that can make judgments or predictions without the need for explicit rule-based programming [104–106]. ML has become a game-changing technique in polymer research, facilitating rapid material discovery [83, 107, 108], synthesis optimization [101, 109–112], and property prediction [113–119]. Supervised learning (SL), unsupervised learning (UL), and reinforcement learning (RL) are the three main paradigms into which machine learning (ML) approaches are divided. These paradigms differ in how they handle data, the degree of human supervision they require, and how they approach problems [120–122]. While unsupervised learning identifies latent patterns in unlabelled data, which are frequently used for clustering or dimensionality reduction, supervised learning uses labelled datasets to train prediction models for tasks like regression or classification. In contrast, reinforcement learning is the process by which agents interact with their environment iteratively to learn the best decision-making procedures. Several easily available resources offer useful, application-focused advice for researchers and practitioners looking to quickly deploy ML techniques. These consist of practical tutorials and chemist-specific executable codebases that need little setup and allow for instant testing and deployment [123]. Prior to exploring specific machine learning paradigms and their corresponding applications in polymer science, it is important to recognise the variety of algorithmic approaches that fall under the purview of machine learning. The effectiveness of deep learning (DL) in processing high-dimensional and complex polymer datasets has attracted a lot of attention, but traditional machine learning (ML) algorithms like Random Forests (RF) [70], Support Vector Machines (SVM) [124], Principal Component Analysis (PCA) [125], and k-Means Clustering [126] are still very useful. These conventional models are preferred for their interpretability and computational efficiency, particularly in situations where model transparency is crucial, and they provide clear benefits in situations involving limited or structured datasets.

Comprehensive studies can be obtained in foundational texts and specialised reviews [127, 128], even though a thorough exposition of the algorithmic structures and theoretical foundations is outside the current purview. Notably, artificial neural networks (ANNs), which are mathematical constructions modelled after the design of biological neural systems, serve as the foundation for DL [94]. As schematically shown in Fig. 4, these networks are made up of several interconnected layers of nodes (neurones) that enable end-to-end learning and hierarchical feature extraction from raw or designed input data. Raw or pre-processed data is ingested by the input layer of a deep neural network architecture, and it undergoes a series of transformations through several hidden layers. The network can learn intricate hierarchical patterns and abstract representations of the data thanks to the artificial neurones that make up

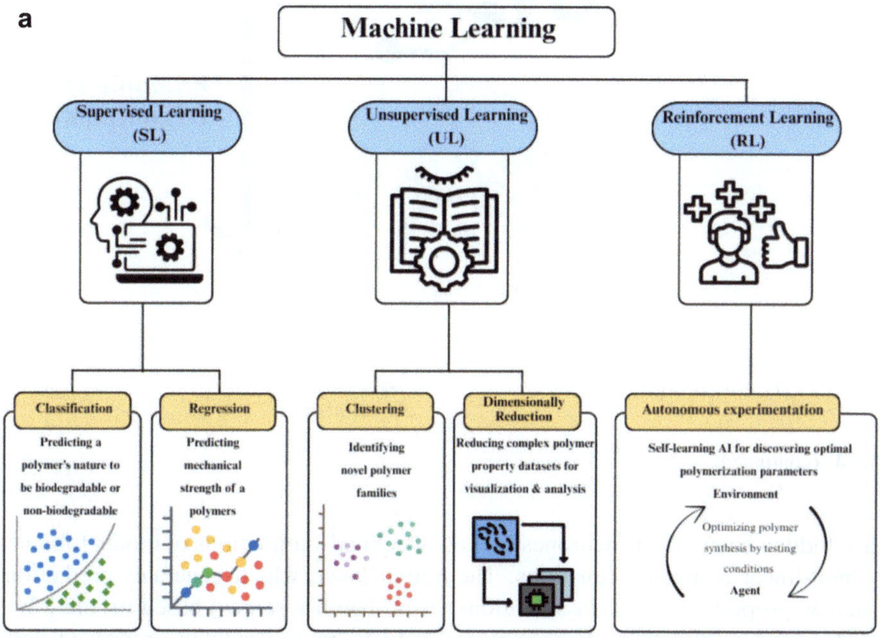

Fig. 4 Overview of main machine learning methods and their applications. Deep learning (DL) can be applied across all three categories to analyse polymer data, predict properties and optimise synthesis. Experimental data source used in ML driven polymer research include atomic force microscopy (AFM), transmission electron microscopy (TEM), and Nuclear magnetic resonance (NMR) microscopy

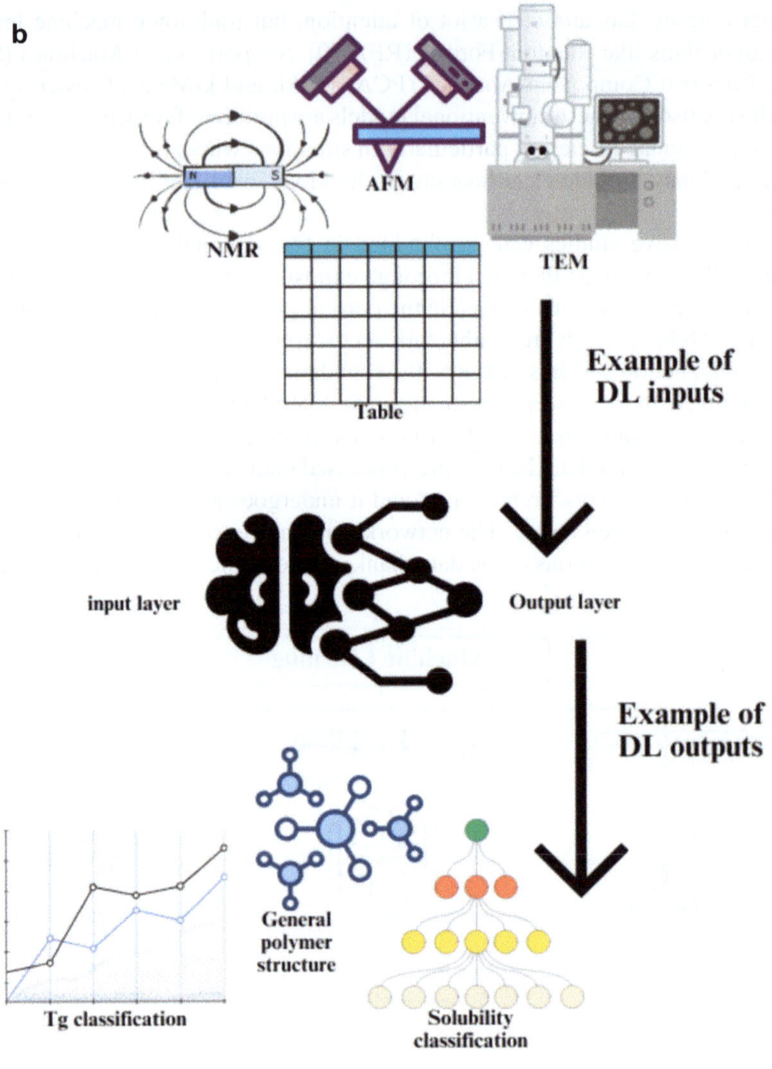

Fig. 4 (continued)

each hidden layer. These neurones execute weighted summations of inputs followed by non-linear activation functions. The output layer, which generates predictions (such as property values, classifications, or probability scores) based on the job at hand, is the culmination of these learnt features being gradually improved upon throughout layers. Under the guidance of a loss function that measures prediction error, optimisation methods like stochastic gradient descent are used to iteratively alter the model's internal parameters (weights and biases) during the learning process. The network improves its predictive accuracy by minimising this loss throughout the course of multiple training epochs. This architecture can be conceptually compared to a multi-tiered analytical system, with the ultimate layer synthesising the information to produce correct outputs, intermediate levels capturing higher-order correlations, and early layers extracting essential features. Deep learning models are able to represent highly nonlinear, multivariate connections because to this multi-layered processing pipeline, which is crucial given the complexity frequently found in polymer systems [94]. Different neural network architectures are designed to handle different kinds of input and analytical tasks. When dealing with structured, tabular data, Fully Connected Neural Networks (FCNNs) [129] are frequently used for classification and regression tasks. By hierarchically extracting spatial features and patterns, Convolutional Neural Networks (CNNs) [130] are excellent at processing images and are optimised for spatial data analysis. Because they are specifically made to handle sequential and temporal data, recurrent neural networks (RNNs) [130], including Long Short-Term Memory (LSTM) networks [131], are useful for time-series forecasting and natural language processing. Several hidden layers make up Deep Neural Networks (DNNs) [132], which are used to describe extremely complicated and nonlinear relationships. Deep learning techniques are being used into polymer informatics processes more frequently due to the intrinsic complexity of polymer characterisation and property prediction. This has a substantial positive impact on computing efficiency and predictive accuracy.

4.1 Supervised Learning

Similar to traditional pedagogy, which involves a teacher guiding students' learning through examples and feedback, supervised learning (SL) uses labelled datasets such that each input correlates to a known outcome. In contrast to human learning, SL models use optimization algorithms to iteratively assess their predictive performance and adjust model parameters until a predetermined accuracy or loss threshold is reached, guaranteeing accurate and consistent predictions. Regression and classification are the two main task categories that SL approaches focus on. Predicting distinct categorical labels—such as differentiating between biodegradable and non-biodegradable polymers—is the task of classification. SL models identify underlying patterns to reliably classify the biodegradability of novel polymer candidates by training on datasets that include chemical structures and related degradation behaviours. Regression tasks, on the other hand, involve forecasting continuous variables, like the glass transition temperature (Tg). The materials discovery process can be accelerated by using SL models to estimate Tg values for previously uncharacterized polymers by understanding the functional correlations between molecular descriptors and thermal properties [120]. With an accuracy of roughly 93%, the trained model then forecasts the compatibility of new polymer–solvent combinations, far outperforming traditional heuristic techniques like the Hildebrand and Hansen solubility parameters. This development has important ramifications for medication delivery, membrane technology, and plastics recycling applications where solvent choice has a major impact on material processing and functional performance. Lu et al. [114] used random forest methods and supervised learning to forecast phase behaviour in polymerization-induced self-assembly (PISA) in a different study. Using a dataset of 592 experimentally characterised samples labelled by observed morphologies (e.g., spheres, worms, and vesicles), the model used characteristics including block ratio, polymerisation conditions, and monomer composition to accurately categorise PISA results. Crucially, the model's intrinsic interpretability made it possible to clarify important chemical characteristics causing phase transitions, providing both predictive and mechanistic insights. Building upon this concept, Fonseca Parra et al. [56] developed a deep learning framework to generate three-dimensional pseudo-phase diagrams for block copolymers. By combining several processing parameters at once, their deep neural network model, which was trained on extensive literature datasets, is able to capture intricate morphological transitions. This method clarifies nonlinear relationships between polymer composition, concentration, and self-assembly behaviour, in contrast to conventional two-dimensional phase diagrams that take into account a small number of factors. As a result, it offers improved predictive power for morphology evolution that goes beyond the limitations of traditional heuristic models. Automating intricate data analysis tasks, particularly in microscopy image processing, has been accomplished with success using supervised learning (SL). The accurate localisation and quantification of nanoparticles inside polymer matrices, which is typically dependent on manual or time-consuming image analysis approaches, is a crucial difficulty in polymer nanocomposite research. Using convolutional neural networks (CNNs) trained on labelled transmission electron microscopy (TEM) datasets, Qu et al. [133] created a deep learning-based method to get around issue. Their process used 72 TEM pictures, from which an automated cropping and annotation methodology (DOPAD) produced 279,057 labelled sub-images. In comparison to traditional human methods, the trained CNN model significantly improved the speed and accuracy of nanoparticle characterisation by predicting nanoparticle positions and sizes in unique TEM images with high accuracy. This development speeds up research in functional polymer nanotechnologies, advanced materials, and coatings by greatly simplifying the study of polymer nanocomposites. When creating supervised learning frameworks, it is essential to distinguish between different input data modalities. Structured chemical descriptors based on representations like SMILES strings or molecular fingerprints are commonly used in property prediction tasks like glass transition temperature (Tg) or solubility estimation. On the other hand, image-based studies, such as atomic force microscopy (AFM) and transmission electron microscopy (TEM), require specialised models that analyse raw pixel-level input, including convolutional neural networks (CNNs) or object detection architectures like YOLOv. While descriptor-based models are extremely sensitive to the selection, quality, and physicochemical significance of input characteristics, image data frequently necessitates substantial manual annotation and strong data augmentation techniques to improve model generalisation. To maximise model accuracy, robustness, and interpretability, these modality-specific features must be effectively recognised and accommodated [133]. On the other hand, image-based studies, such as atomic force microscopy

(AFM) and transmission electron microscopy (TEM), require specialised models that analyse raw pixel-level input, including convolutional neural networks (CNNs) or object detection architectures like YOLOv. While descriptor-based models are extremely sensitive to the selection, quality, and physicochemical significance of input characteristics, image data frequently necessitates substantial manual annotation and strong data augmentation techniques to improve model generalisation. To maximise model accuracy, robustness, and interpretability, these modality-specific features must be effectively recognised and accommodated [133].

4.2 Unsupervised Learning

When no output labels are given, unsupervised learning (UL) is a reliable technique for identifying underlying patterns in unlabelled datasets [121]. UL independently investigates data structures to find latent groups, correlations, or hierarchical relationships, in contrast to supervised learning, which relies on matched input–output data. From a conceptual standpoint, UL functions similarly to an autonomous learner deriving thematic insights from textual data without outside guidance. Because of this feature, UL is especially useful for examining intricate polymer datasets with little or no experimental annotations. UL approaches are widely used for dimensionality reduction, which reduces high-dimensional data while maintaining important informational content, and for grouping polymers based on shared chemical or structural characteristics [134, 135]. They were able to identify different micelle populations with similar corona configurations by clustering molecular conformations, and they were able to simplify complex structural variability by dimensionality reduction. This methodology clarified basic self-assembly mechanisms that are essential for drug delivery and biomaterials engineering applications. They were able to identify different micelle populations with similar corona configurations by clustering molecular conformations, and they were able to simplify complex structural variability by dimensionality reduction. This methodology clarified basic self-assembly mechanisms that are essential for drug delivery and biomaterials engineering applications. They were able to identify different micelle populations with similar corona configurations by clustering molecular conformations, and they were able to simplify complex structural variability by dimensionality reduction. This methodology clarified basic self-assembly mechanisms that are essential for drug delivery and biomaterials engineering applications. Without the requirement for previous labelling, this method allowed polyolefin samples to be grouped according to spectral fingerprint similarities. The resulting lower-dimensional form made it easier to visualise and comprehend the dataset, exposing structural traits and polymer composition-correlated groups having chemical significance. Thus, UL through fPCA revealed hidden chemical patterns that conventional analytical techniques would probably miss, proving its usefulness in spectral data analysis for polymer characterisation [134, 135].

4.3 Reinforcement Learning and Closed-Loop Optimization

Reinforcement Learning (RL) is a unique machine learning paradigm in which agents interact iteratively with a predefined environment to learn optimal policies, and they get scalar reward signals when they do advantageous behaviours [122]. Unlike supervised learning, which uses labelled input–output pairs, reinforcement learning (RL) is based on experiential learning, which is similar to an agent learning from direct feedback from its surroundings instead of from previously marked examples. RL frameworks need sequential decision-making, cumulative reward maximisation, and the handling of the exploration–exploitation issue, they are fundamentally more complicated than supervised or unsupervised learning. To converge on an ideal strategy, agents must weigh the trade-off between investigating novel tactics and taking use of well-known rewarding behaviours. RL adjusts dynamically, improving its decision-making approach in response to changing environmental feedback, in contrast to static data-driven models. Because of these features, RL is especially useful for polymer science applications involving dynamic control or optimisation under uncertainty. Closed-loop experimental design, real-time polymerisation parameter adjustment, and autonomous robotic experimentation are noteworthy examples [101, 136]. These advantages, however, come at the expense of greater computational and algorithmic complexity, frequently requiring sophisticated designs (such as actor-critic approaches, policy gradients, and Q-learning) and substantial computer resources for successful implementation. In contrast to traditional static reaction protocols, Li et al. [136] used a reinforcement learning (RL)-driven approach to control molecular weight distribution (MWD) in atom transfer radical polymerisation (ATRP). The polymerisation reactor serves as the interactive environment in their system, which incorporates an AI agent—which consists of policy and value networks—into a real-time feedback loop. The agent is rewarded according to how closely the generated MWD resembles a predetermined target distribution (e.g., Gaussian or bimodal profiles) and chooses reagent addition strategies based on the observed reaction state variables, such as monomer and initiator concentrations. By comparing expected and target MWDs, the model continuously improves its action strategy, allowing for dynamic optimisation of reaction circumstances through iterative exploration. The precision of regulating MWD is much improved by this closed-loop RL method, which outperforms heuristic-based techniques. It demonstrates the promise of reinforcement learning (RL) for autonomous process control and adaptive materials design in polymer science by making it easier to synthesise custom polymer architectures while reducing the number of experimental iterations. Practical difficulties still limit the use of reinforcement learning (RL) in polymer research, despite its potential. RL frameworks usually require either access to high-fidelity simulation environments or substantial real-time experimentation, both of which are resource and computationally demanding. Furthermore, it is frequently difficult to formulate suitable reward functions and action spaces for complicated polymer systems, which restricts RL's applicability to wider exploratory domains. As a result, RL is now more appropriate for optimisation tasks with limited scope, like optimising certain polymerisation processes [122]. Hybrid approaches that combine reinforcement learning (RL) with supervised learning (SL) or Bayesian optimisation (BO) are becoming viable substitutes to address these drawbacks. For example, Pittaway et al. [137] showed how to enable closed-loop self-optimization of emulsion polymerisation in a continuous-flow reactor by integrating multi-objective BO with real-time dynamic light scattering (DLS) feedback. With little assistance from humans, this technique allowed for the autonomous adjustment of reaction conditions to maximise a variety of goals, including particle size and dispersion. An AI-guided closed-loop system for reversible addition–fragmentation chain-transfer (RAFT) polymerisation optimisation was also created by Warren et al. [101]. Using the Thompson Sampling Efficient Multi-Objective Optimisation (TSEMO) method, their software integrated real-time analytical feedback from nuclear magnetic resonance (NMR) and gel permeation chromatography (GPC) inside a Bayesian optimisation framework (Fig. 5). With few experimental repetitions, the system effectively reached desired molecule weights and dispersities by independently exploring reaction condition space and iteratively improving its approach. These developments highlight the increasing potential of hybrid AI techniques to speed up polymer process optimisation while preserving experimental effectiveness.

The Warren et al. [101] platform does not rely on pre-established datasets; instead, it learns from its own experimental results and adaptively adjusts parameters like temperature and reaction time to maximise monomer conversion and minimise dispersity. Through an active learning loop, the system uses a predictive surrogate model to estimate the predicted outcomes of untested reaction conditions, allowing for well-informed experimental selection. In order to achieve effective multi-objective optimisation, our model-guided exploration strikes a compromise between exploring uncertain portions of the reaction space and taking advantage of high-performing conditions. This approach demonstrates the effectiveness of autonomous experimentation and adaptive control in polymer synthesis, while not being a pure reinforcement learning (RL) framework. The platform may iteratively improve its comprehension of the reaction landscape by combining Bayesian optimisation with real-time analytical input. This self-directed method speeds up process optimisation, improves reaction accuracy, and requires less human intervention. The paper highlights a paradigm shift in polymer chemistry towards semi-autonomous, self-learning systems, opening the door for closed-loop material discovery platforms powered by artificial intelligence. These illustrated case studies highlight the diverse applications of machine learning (ML) in polymer research, including data-driven discovery workflows, autonomous process optimisation, and predictive modelling of polymer properties. Depending on the problem setting, each of the machine learning (ML) paradigms—supervised learning (SL), unsupervised learning (UL), and reinforcement learning (RL)—offers distinct methodological benefits. UL uncovers latent structures in high-dimensional data without labels [108], RL enables adaptive decision-making in dynamic contexts like real-time polymerisation control [136], and SL performs exceptionally well in regression and classification tasks using well-annotated datasets [114, 136]. A comparison of these methods is given

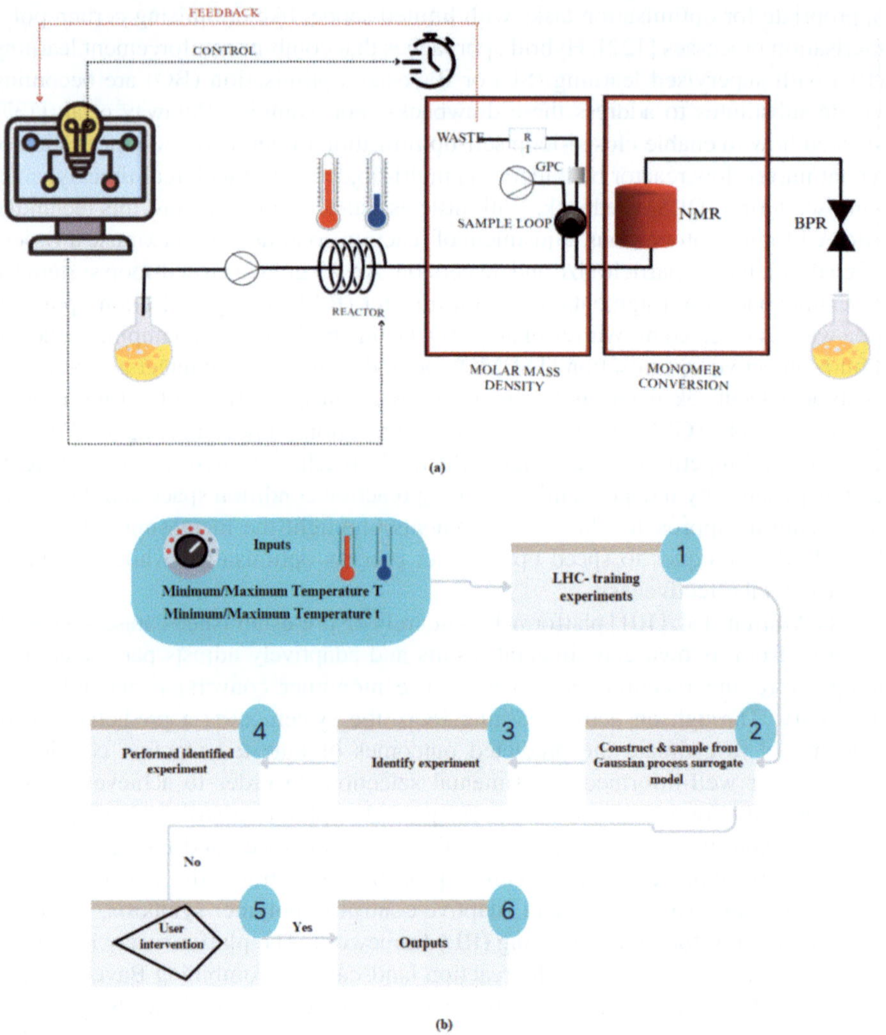

Fig. 5 **a** AI-guided closed-loop optimisation of reversible addition-fragmentation chain transfer (RAFT) polymerisation using Bayesian optimisation. **b** An overview of the structure of the Thompson-sampling efficient multi-objective optimisation (TSEMO) algorithm-based experiments

in Table 2, which also outlines the data requirements, optimisation frameworks, and pertinent areas of polymer research for each. Crucially, they differ greatly in terms of interpretability and computing expense, thus careful selection is required depending on the goals and complexity of a particular activity.

Even while ML-driven methods are becoming more and more popular, it's important to understand their limitations. Rule-based or programmatic methods may provide more effective and comprehensible solutions than sophisticated machine

Table 2 Comparison of key machine learning approaches

Feature	Supervised learning (SL)	Unsupervised learning (UL)	Reinforcement learning (RL)
Data type	Data with labels (input–output pairs)	Unlabeled data (identifying trends)	Learn from interactions and has no predetermined labels
Goal	Forecast results (regression/ classification)	Reduce dimensions or organize related data into clusters	Discover a series of steps to increase rewards
Learning process	Learns from explicit examples	Finds unseen structures on its own	Learns by making mistakes and getting feedback from the environment
Optimization focus	Reduce loss (error)	Examine groups, trends, and representations	Increase long-term benefits
Computational complexity	Moderate	Moderate to high	Very high (complex decision-making)

learning algorithms in systems with low dimensionality or well-understood mechanical behaviour. Therefore, to guarantee the successful integration of ML into polymer research workflows, careful model selection, thorough benchmarking, and alignment with the experimental design space are essential.

5 Real-World ML Tools

A wide range of artificial intelligence (AI) tools and platforms that support data collection, processing, modeling, and deployment significantly ease the integration of machine learning (ML) into polymer research. These resources cover a broad range of functions, such as cloud-based execution environments (like Google Colab and Jupyter Notebooks), no-code machine learning platforms (like KNIME and DataRobot), visualization frameworks (like Matplotlib and Seaborn), cheminformatics toolkits (like RDKit and DeepChem), and data manipulation libraries (like Pandas and NumPy. Moreover, curated datasets designed for polymer informatics applications are available through polymer-specific repositories and databases as PoLyInfo, Polymer Genome, and the Materials Project. The creation of prediction models for polymer performance, compatibility, and characteristics is made possible by these platforms. In this context, open-source software is essential because it fosters accessibility, repeatability, and transparency. These tools democratize AI-driven research by reducing the entry hurdle for non-experts, hastening the adoption of data-centric approaches in polymer science and encouraging cross-disciplinary collaboration. For machine learning (ML) in polymer science, Python has become the most popular programming language because of its user-friendly syntax, cross-platform portability, and extensive collection of scientific libraries. Without the need for sophisticated setup or expensive computing infrastructure, interactive development environments like Google Colab and Jupyter Notebooks offer easily accessible platforms for code execution, data analysis, and model deployment. Plots of publication quality and exploratory data analysis are supported by visualization packages like Matplotlib and Seaborn, while core Python libraries like NumPy and Pandas enable effective numerical computations and data preprocessing. Tasks like feature engineering, cleaning, loading datasets, and evaluating models can be completed more quickly thanks to these tools. Low-code and no-code machine learning platforms are good options for researchers who don't know much about programming. Using drag-and-drop modules, tools such as KNIME offer graphical interfaces for creating end-to-end machine learning pipelines that include data transformation, method selection, and model validation. With user-friendly interfaces designed for quick prototyping and categorization tasks, platforms like Google AutoML, Azure ML Studio, and Google's Teachable Machine further streamline model training and deployment. A larger community of polymer scientists may now use machine learning (ML) thanks to these resources, which drastically reduce the entry barrier. Using cheminformatics toolkits to convert molecular representations into machine-interpretable formats is a crucial use of machine learning (ML) in polymer research. A popular open-source tool called RDKit makes it possible to transform chemical structures into formats like molecular fingerprints and SMILES strings, which are used as standardized inputs for machine learning models. By incorporating stochastic and repeating aspects of polymer chains, the BigSMILES notation expands this functionality to address the structural variability of macromolecular systems and enhances model applicability to genuine polymeric materials. By providing pre-trained machine learning models that can predict a variety of polymer properties (such as the dielectric constant and glass transition temperature) straight from structural descriptors, Polymer Genome further improves this paradigm and drastically cuts down on the time needed for materials discovery and candidate screening. Another key benefit of ML integration is its capacity to effectively handle and analyze high-dimensional polymer datasets. Beyond the scalability of traditional tools like Microsoft Excel, libraries like Pandas enable sophisticated data processing operations—such as filtering polymers based on molecular weight thresholds, calculating property correlations, or producing descriptive statistics—in a matter of seconds. Faster hypothesis testing, improved repeatability, and data-driven experimental design are all made possible by these optimized methods. As open-source platforms become more widely available, it is getting easier to incorporate machine learning into polymer research. Robust predictive modeling with low overhead is made possible by beginner-friendly frameworks like Scikit-learn, and automated data pipelines can be supported by graphical workflow tools like KNIME without the need for substantial coding knowledge. Frameworks like TensorFlow and PyTorch provide scalable architectures appropriate for complex model construction and polymer property prediction for sophisticated deep learning applications. A summary of the key platforms and innovations for machine learning in polymer science is shown in Table 3.

Table 3 A summary of the key platforms and innovations for machine learning in polymer science. P: proprietary, FT: free-tier, and OS: open-source

Category	Programms	Functionality
Low-code/no-code machine learning platforms	Google AutoML (P), Azure ML (P), Weka (OS), KNIME (OS), and teachable machines (FT)	ML without scripting through graphical user interfaces; perfect for basic processes, clustering, and classification
Platforms for sharing and collaboration	GitHub (OS), Zenodo (OS), Hugging Face (OS), and Figshare (OS)	Code, datasets, and trained models are hosted; version control and DOI-based citation are supported
Environments for programming and execution	Google Colab (FT), Jupyter Notebooks (OS), Anaconda (FT), and Python (OS)	Data science workflows involving interactive coding, script execution, and environment management
Data preparation and manipulation	Pandas (OS) and Numpy (OS)	Effective management of arrays, tables, and structured experimental data
Visualization of data	Seaborn (OS) and Matplotlib (OS)	Data and model outputs are graphically represented for analysis and communication
Machine learning libraries	TensorFlow (OS), PyTorch (OS), and Scikit-learn (OS)	Libraries for deep learning and traditional machine learning, including neural networks, regression, and categorisation
Descriptors and chemical representation	RDKit (OS), BigSMILES (OS), and SMILES	Creation of chemical descriptors and encoding of molecular and polymeric structures
Polymer data repositories	PoLyInfo, PI1M, CROW, NIST DB (R/FT), Materials Project (FT), and Polymer Genome (FT)	Databases of polymer characteristics from computer and experiment

6 Challenges and Considerations

Although artificial intelligence (AI) has the potential to revolutionise polymer research, a number of enduring obstacles continue to impede its practical application. The increased interest in machine learning (ML) applications is demonstrated by the expanding number of AI-driven papers, but this excitement has not yet resulted in widespread acceptance in industrial settings and experimental labs. The lack of standardised processes for dataset curation and sharing, inconsistent data quality, and restrictions on data availability are major obstacles. The creation of generalisable models is hampered by the sparsity, variability, and lack of reproducibility that plague polymer datasets. Additionally, polymer chemists and materials scientists without computational backgrounds face a substantial entry hurdle due to the learning curve associated with machine learning tools and programming languages. Access to high-performance infrastructures needed for training and implementing complicated machine learning models, like deep neural networks, is further hampered by computational limitations, especially in academic settings with limited resources. Furthermore, the repeatability, scalability, and interpretability of results are hampered by the absence of standardised pipelines for combining machine learning with experimental processes. The creation of easily accessible teaching materials, standardised frameworks for data and model interoperability, and user-friendly machine learning platforms are crucial to facilitating the shift of machine learning from theoretical promise to practical application. Bridging the gap between AI capabilities and their practical use in polymer research would need strategic efforts in open science, interdisciplinary collaboration, and infrastructure development.

7 Data Resources: Availability, Accessibility, and Challenges

Access to high-fidelity, organised datasets and collaborative coding infrastructures are essential for the effective integration of machine learning (ML) into polymer research. By providing curated polymer databases, standardised machine-readable representations, and model sharing repositories, a number of projects have arisen to meet this demand and make model training, validation, and reproducibility easier. The NIST Polymer Database, a collection of experimentally verified polymer properties that acts as a standard for model evaluation, the Materials Project, which offers computationally derived data pertinent to polymeric systems, and Polymer Genome, which uses machine learning (ML) to predict properties across a broad range of polymer classes are important platforms. Although there is still a lack of cross-platform interoperability and integration, other domain-specific resources like PoLyInfo and PI1M provide organised datasets including chemical structures and physical attributes. The majority of polymer science datasets are still in the "small data" category, often consisting of tens to hundreds of hand selected items drawn from diverse literature, even with the increasing number of tools available. This poses a serious obstacle to the creation of reliable and broadly applicable machine learning models, especially for those that use deep learning architectures that demand substantial amounts of training data. Standardised data schemas, community-driven data generating initiatives, and easier access to high-throughput experimental and simulation datasets are all necessary to overcome this limitation. Enabling scalable and dependable AI-driven innovation in polymer design and discovery requires several actions. A number of general-purpose platforms facilitate organised data sharing, model distribution, and collaborative coding in addition to polymer-specific repositories, offering important infrastructure for developing AI applications in polymer science. A wide range of pre-trained machine learning models, including those tailored for polymer-related tasks, are hosted on platforms like Hugging Face, enabling quick model deployment and community benchmarking. Data transparency and academic credit are promoted by Zenodo, a publicly available repository that makes it possible to save datasets and trained models with appropriate citation via Digital Object Identifiers (DOIs). Reproducibility and methodological rigour in polymer informatics are reinforced by GitHub, which is still essential for collaborative development, open-source code sharing, and version-controlled AI pipelines. The sector still has difficulties with data standardisation, completeness, and accessibility despite the expanding ecosystem of these technologies. The majority of current research is based on disjointed, irregularly organised datasets that frequently lack important metadata about the pathways of polymer synthesis, processing settings, and characterisation procedures. These restrictions make machine learning models less reproducible and less broadly applicable in many research settings.

On the other hand, fields like materials science and systems biology have advanced significantly through the adoption of standardised ontologies and community-wide data sharing procedures. Enforced data-sharing standards are desperately needed to keep polymer science from falling behind. Deposition of organised, annotated datasets should be a requirement for publishing or grant payment from funding bodies and scientific publications. In addition to enhancing data quality, this will hasten the creation of high-volume, interoperable datasets that are necessary for reliable and broadly applicable machine learning models. To fully realise the potential of AI-driven polymer research, common processes for data gathering, annotation, and distribution must be established. A methodical and standardised strategy to data preprocessing and curation is necessary to convert raw experimental data into formats that are suited for machine learning (ML). It is recommended that researchers use the following organised workflow: (i) consistently encode macromolecular structures using standardised polymer representations like SMILES or BigSMILES, (ii) add detailed metadata to datasets, such as synthesis parameters, processing conditions, and characterisation methods, (iii) perform necessary data cleaning—removing missing values, duplicates, and outliers—and (iv) distribute carefully selected datasets via open-access sites like Zenodo, GitHub, or domain-specific repositories like Polymer Genome. Maintaining transparency and reproducibility in machine learning pipelines requires that datasets be integrated with version control systems and stored in machine-readable forms (such as CSV, JSON, and HDF5). In addition to sharing data, researchers are highly urged to share pre-trained models when possible and publish related ML codebases and methods. These procedures improve reproducibility, encourage model generalisability across applications, and drastically cut down on redundancy. For both datasets and computational artefacts, it is important to prioritise adherence to the FAIR principles: Findability, Accessibility, Interoperability, and Reusability. The polymer science community can fully utilise ML techniques when FAIR-aligned data infrastructures are put into place since they encourage collaborative research. The field may advance polymer informatics and AI-assisted materials discovery more quickly by moving towards open scientific principles, which will guarantee that curated data resources are not only broadly available but also interoperable and effectively used across research fields.

8 Polymer Science Education Gaps: The Necessity of Interdisciplinary Approaches

For academics who have historically relied on empirical investigation and traditional computational modelling, the incorporation of artificial intelligence (AI) into polymer science represents a paradigm shift. A durable change requires that AI and machine learning (ML) education be integrated directly into polymer science curricula, even though temporary multidisciplinary partnerships between polymer scientists and AI experts provide a first step. Training computer scientists in the complexities of polymer chemistry and engineering is typically less practical than providing polymer researchers with fundamental AI/ML skills, given the domain-specific complexity of polymer synthesis, processing, and characterisation. However, there is still a lack of formalised training in data-driven approaches in polymer science programs, despite AI's increasing impact on materials discovery and design. There is a dearth of researchers skilled in both fields since few master's or doctoral programs provide specialised coursework or practical experience that combines polymer science with computational intelligence. Since the polymer industry is looking for people who can use AI to speed material research, optimise processes, and conduct intelligent experiments, this educational gap not only hinders academic innovation but also industrial advancement. The creation of integrated curriculum that incorporate data science, machine learning, and polymer chemistry is necessary to close this gap. Access to multidisciplinary research opportunities and open-source instructional materials is also necessary. The development of a new generation of researchers capable of spearheading the advancement of AI-assisted polymer science depends on such initiatives. Numerous prominent industrial organisations have already initiated the integration of AI-powered approaches into their R&D processes. While Dow Chemical uses machine learning (ML) techniques for process optimisation, BASF has made significant investments in AI to speed up materials discovery. Arkema has started AI-based initiatives for creative material development, and Covestro has embraced AI to promote sustainable polymer design. Despite these developments, the polymer industry's wider potential for AI is still largely untapped, mainly because there aren't many experts who can bridge the gap between data science and polymer engineering. Incorporating AI, ML, and data science courses specifically designed for polymer science into university curricula is necessary to bridge this skills gap. Emerging polymer scientists will have the skills necessary to smoothly incorporate computational techniques, algorithmic modelling, and AI toolkits into their research workflows if they receive early pedagogical exposure. Furthermore, academic and industry researchers will be better equipped to embrace AI methodologies with more assurance and independence if the current workforce is upskilled through focused efforts like multidisciplinary seminars, summer schools, and open-access online training platforms. By making AI a fundamental ability integrated into polymer science education and practice rather than the purview of a small number of interdisciplinary specialists, such educational approaches are essential to democratising AI adoption across the polymer domain.

9 Computational Costs

Significant computer resources are required for the integration of AI into polymer research, especially for data-intensive methods like deep learning and high-dimensional chemical modelling. Access to high-performance computing (HPC) infrastructure is a major bottleneck for many labs because tasks like deep neural network training and the analysis of intricate datasets from molecular dynamics simulations or spectroscopic measurements frequently surpass the capabilities of standard computing environments. Numerous government-led programs have surfaced worldwide to address these constraints, providing academics with state-of-the-art HPC resources designed for AI-driven scientific research. Advanced computational research across disciplines is made possible in France via the Grand Équipement National de Calcul Intensif (GENCI), which offers AI-optimized supercomputing resources like Jean Zay. World-class facilities like LUMI in Finland and MeluXina in Luxembourg are supported by the EuroHPC Joint Undertaking at the European level. These facilities were specifically created to speed up data-intensive research in fields like materials and polymer science. Two of the world's most potent supercomputers, Summit and Frontier, are run by the Department of Energy (DOE) in the United States and are made available to the scientific community through joint projects with national labs and universities. Fugaku, a world-renowned HPC platform that supports interdisciplinary research, including polymer informatics and AI-enhanced materials design, is managed by Japan's RIKEN Centre for Computational Science in Asia. Rapid advancements in computational materials science have been facilitated by China's concurrent development of vast AI-focused supercomputing resources in centres like Tianjin and Shenzhen. When taken as a whole, these infrastructures are essential for democratising access to AI capabilities and making it possible for machine learning to be applied scaled in polymer research.

10 Future Perspectives

Data-driven design paradigms, autonomous experimentation, and predictive modeling are all seeing revolutionary breakthroughs as a result of the fusion of artificial intelligence (AI) and polymer science. The focus is moving from simply speeding up polymer discovery to algorithmically guiding it as AI techniques advance. In order to optimize polymerization processes and clarify intricate structure–property interactions, reinforcement learning and unsupervised learning are expected to be crucial. This shift represents a fundamental reorganization of the research process, moving away from empirical trial-and-error and toward the development of hypotheses based on pattern recognition and statistical inference. This paradigm change is best illustrated by AI-driven autonomous laboratories that combine closed-loop learning algorithms, robotic synthesis, and in situ characterisation. By adjusting reaction conditions in real-time, these devices can improve throughput and reproducibility while optimizing molecular designs with little assistance from humans. Intelligent decision-making and automation are two aspects of such platforms' transformational potential. The function of the polymer scientist will change from carrying out experiments to strategically designing data-centric processes as machine learning models become more and more integrated into laboratory operations. Beyond discovery, artificial intelligence will be essential to the development of sustainable polymers. Energy-efficient synthesis route modeling, recycling strategy optimization, and the logical design of biodegradable polymers are all made possible by machine learning approaches. These qualities will be crucial in facilitating the shift to a circular polymer economy, establishing AI as a key force behind the innovation and sustainability of next-generation polymers.

11 Conclusion

With its unparalleled potential to speed up discovery, improve material characteristics, and promote sustainable design, artificial intelligence (AI) has evolved from a cutting-edge auxiliary to a strategic necessity in polymer science. However, polymer scientists must actively participate in order to fully realize AI's transformational promise; data science experts alone won't be enough. This viewpoint emphasizes how urgent it is to incorporate AI as a fundamental part of processes for polymer research. Instead of replacing conventional techniques, machine learning should be seen as an adjunct that improves mechanistic understanding, experimental efficiency, and predictive accuracy. Institutions must prioritize AI teaching in their curriculum, encourage the creation of standardized, machine-readable datasets, and support ongoing interdisciplinary collaboration in order to guarantee widespread and equitable advantages. Despite the obstacles that still exist, such as the lack of data and the high learning curve of AI tools, incremental adoption is both possible and effective. Every fundamental step creates the framework for a digitally enhanced research culture, whether it is sharing curated data, training a basic model, or integrating AI into synthesis processes. The role of the polymer scientist will change toward precision-guided material design as AI-powered labs automate and improve routine testing, opening a new era where data-driven insights outweigh traditional trial-and-error methods.

Acknowledgements Authors are thankful to Director, Amity Institute of Applied Sciences (AIAS), and Head, Department of Chemistry for providing the facilities. One of the authors, Tejendra K Gupta is thankful to the Council of Science & Technology, Lucknow, Uttar Pradesh, India for the financial support (Vide No. CST/CHEM/D-1431).

References

1. Pfleger, J., Cimrová, V.: Polymers for Electronics and Photonics: science for Applications. Springer International Publishing (2018). https://doi.org/10.1007/s11696-018-0472-0
2. Huan, T.D., et al.: Advanced polymeric dielectrics for high energy density applications (Elsevier Ltd.) (2016). https://doi.org/10.1016/j.pmatsci.2016.05.001
3. Sionkowska, A.: Current research on the blends of natural and synthetic polymers as new biomaterials: review (2011). https://doi.org/10.1016/j.progpolymsci.2011.05.003
4. Alam, A.U., et al.: Polymers and organic materials-based pH sensors for healthcare applications (Elsevier Ltd.) (2018). https://doi.org/10.1016/j.pmatsci.2018.03.008.
5. Liles, D.T., Lin, F.: Silicone elastomeric particles in skin care applications. In: ACS Symposium Series, American Chemical Society, pp. 207–219 (2010). https://doi.org/10.1021/bk-2010-1053.ch011
6. Nandi, A.K., Chatterjee, D.P.: Hybrid polymer gels for energy applications. Royal Society of Chemistry (2023). https://doi.org/10.1039/d2ta09525d.
7. Zhang, G., Cheng, X., Wang, Y., Zhang, W.: Supramolecular chiral polymeric aggregates: construction and applications (Wiley) (2023). https://doi.org/10.1002/agt2.262.
8. Geyer, R., Jambeck, J.R., Law, K.L.: Production, use, and fate of all plastics ever made (2017). https://www.science.org
9. Fredrickson, G.H., et al.: Ionic compatibilization of polymers. ACS Polym. Au **2**(5), 299–312 (2022). https://doi.org/10.1021/acspolymersau.2c00026
10. Kumar, S.K., Krishnamoorti, R.: Nanocomposites: structure, phase behavior, and properties. Annu. Rev. Chem. Biomol. Eng. **1**, 37–58 (2010). https://doi.org/10.1146/annurev-chembioeng-073009-100856
11. Sarkar, B., Alexandridis, P.: Block copolymer-nanoparticle composites: structure, functional properties, and processing (Elsevier Ltd.) (2015). https://doi.org/10.1016/j.progpolymsci.2014.10.009
12. Correa-Baena, J.P., et al.: Accelerating materials development via automation, machine learning, and high-performance computing. Cell Press. https://doi.org/10.1016/j.joule.2018.05.009
13. Sha, W., et al.: Artificial intelligence to power the future of materials science and engineering. Adv. Intell. Syst. **2**(4) (2020). https://doi.org/10.1002/aisy.201900143
14. Kalidindi, S.R.: Feature engineering of material structure for AI-based materials knowledge systems. J. Appl. Phys. **128**(4) (2020). https://doi.org/10.1063/5.0011258
15. Choudhary, K., Bercx, M., Jiang, J., Pachter, R., Lamoen, D., Tavazza, F.: Accelerated discovery of efficient solar cell materials using quantum and machine-learning methods. Chem. Mater. **31**(15), 5900–5908 (2019). https://doi.org/10.1021/acs.chemmater.9b02166
16. Hattrick-Simpers, J.R., Choudhary, K., Corgnale, C.: A simple constrained machine learning model for predicting high-pressure-hydrogen-compressor materials. Mol. Syst. Des. Eng. **3**(3), 509–517 (2018). https://doi.org/10.1039/c8me00005k
17. Chun, S., Roy, S., Nguyen, Y.T., Choi, J.B., Udaykumar, H.S., Baek, S.S.: Deep learning for synthetic microstructure generation in a materials-by-design framework for heterogeneous energetic materials. Sci. Rep. **10**(1) (2020). https://doi.org/10.1038/s41598-020-70149-0
18. Guo, K., Yang, Z., Yu, C.H., Buehler, M.J.: Artificial intelligence and machine learning in design of mechanical materials. R. Soc. Chem. (2021). https://doi.org/10.1039/d0mh01451f
19. Shen, S.C., Buehler, M.J.: Nature-inspired architected materials using unsupervised deep learning. Commun. Eng. **1**(1) (2022). https://doi.org/10.1038/s44172-022-00037-0
20. Gu, G.X., Chen, C.T., Richmond, D.J., Buehler, M.J.: Bioinspired hierarchical composite design using machine learning: simulation, additive manufacturing, and experiment. Mater. Horiz. **5**(5), 939–945 (2018). https://doi.org/10.1039/c8mh00653a
21. Zhang, G., et al.: Reinforced concrete deep beam shear strength capacity modelling using an integrative bio-inspired algorithm with an artificial intelligence model. Eng. Comput. **38**, 15–28 (2022). https://doi.org/10.1007/s00366-020-01137-1

22. Ishiwatari, Y., Yokoyama, T., Kojima, T., Banno, T., Arai, N.: Machine learning prediction of self-assembly and analysis of molecular structure dependence on the critical packing parameter. Mol. Syst. Des. Eng. **9**(1), 20–28 (2023). https://doi.org/10.1039/d3me00151b
23. Wei, J., et al.: Machine Learning in Materials Science. Blackwell Publishing Ltd. (2019). https://doi.org/10.1002/inf2.12028
24. Zhou, T., Song, Z., Sundmacher, K.: Big data creates new opportunities for materials research: a review on methods and applications of machine learning for materials design. Elsevier Ltd. (2019). https://doi.org/10.1016/j.eng.2019.02.011
25. Goswami, L., Deka, M.K., Roy, M.: Artificial intelligence in material engineering: a review on applications of artificial intelligence in material engineering. Wiley (2023) https://doi.org/10.1002/adem.202300104
26. Yu, C.H., Qin, Z., Buehler, M.J.: Artificial intelligence design algorithm for nanocomposites optimized for shear crack resistance. Nano Futures **3**(3) (2019). https://doi.org/10.1088/2399-1984/ab36f0
27. Gu, G.H., Noh, J., Kim, I., Jung, Y.: Machine learning for renewable energy materials. Royal Society of Chemistry (2019). https://doi.org/10.1039/c9ta02356a
28. Yang, K., et al.: Predicting the young's modulus of silicate glasses using high-throughput molecular dynamics simulations and machine learning. Sci. Rep. **9**(1) (2019). https://doi.org/10.1038/s41598-019-45344-3
29. Liao, M., et al.: Modeling of alloying effect on elastic properties in BCC Nb-Ti-V-Zr solid solution: from unary to quaternary. Comput. Mater. Sci. **172** (2020). https://doi.org/10.1016/j.commatsci.2019.109289
30. Venkatraman, V., Evjen, S., Knuutila, H.K., Fiksdahl, A., Alsberg, B.K.: Predicting ionic liquid melting points using machine learning. J. Mol. Liq. **264**, 318–326 (2018). https://doi.org/10.1016/j.molliq.2018.03.090
31. Chen, L., Tran, H., Batra, R., Kim, C., Ramprasad, R.: Machine learning models for the lattice thermal conductivity prediction of inorganic materials (2019). http://arxiv.org/abs/1906.06378
32. Luo, Y., Li, M., Yuan, H., Liu, H., Fang, Y.: Predicting lattice thermal conductivity via machine learning: a mini review. Nat. Res. (2023) https://doi.org/10.1038/s41524-023-00964-2
33. Chen, Z., et al.: Machine learning on neutron and x-ray scattering and spectroscopies. Chem. Phys. Rev. **2**(3) (2021). https://doi.org/10.1063/5.0049111
34. Deagen, M.E., et al.: Networks and interfaces as catalysts for polymer materials innovation. Cell Press (2022). https://doi.org/10.1016/j.xcrp.2022.101126
35. Meijer, H.E.H., Govaert, L.E.: Mechanical performance of polymer systems: the relation between structure and properties. In: Progress in Polymer Science (Oxford), pp. 915–938 (2005). https://doi.org/10.1016/j.progpolymsci.2005.06.009
36. Lutz, J.F., Lehn, J.M., Meijer, E.W., Matyjaszewski, K.: From precision polymers to complex materials and systems. Nature Publishing Group (2016). https://doi.org/10.1038/natrevmats.2016.24
37. Gurnani, R., Kuenneth, C., Toland, A., Ramprasad, R.: Polymer informatics at scale with multitask graph neural networks. Chem. Mater. **35**(4), 1560–1567 (2023). https://doi.org/10.1021/acs.chemmater.2c02991
38. Patra, T.K.: Data-driven methods for accelerating polymer design. ACS Polym. Au **2**(1), 8–26 (2022). https://doi.org/10.1021/acspolymersau.1c00035
39. Egrioglu, E., Bas, E.: A new deep neural network for forecasting: deep dendritic artificial neural network. Artif. Intell. Rev. **57**(7) (2024). https://doi.org/10.1007/s10462-024-10790-7
40. Zhou, J., et al.: Graph neural networks: a review of methods and applications. KeAi Communications Co. (2020). https://doi.org/10.1016/j.aiopen.2021.01.001
41. Martin, T.B., Audus, D.J.: Emerging trends in machine learning: a polymer perspective. Am. Chem. Soc. (2023). https://doi.org/10.1021/acspolymersau.2c00053
42. Schmid, F.: Understanding and modeling polymers: the challenge of multiple scales. Am. Chem. Soc. (2023). https://doi.org/10.1021/acspolymersau.2c00049
43. Ignacz, G., et al.: Machine learning for the advancement of membrane science and technology: a critical review. Elsevier B.V. (2025). https://doi.org/10.1016/j.memsci.2024.123256

44. Chen, L., et al.: Polymer informatics: current status and critical next steps. Elsevier Ltd. (2021). https://doi.org/10.1016/j.mser.2020.100595
45. Sha, W., et al.: Machine learning in polymer informatics. Blackwell Publishing Ltd. (2021). https://doi.org/10.1002/inf2.12167
46. Khatoon, N., Ali, N., Ali, S., Chen, Z., Jun, W., Yang, H.: Preparation of a CPVC composite loose nanofiltration membrane based on plant polyphenols for effective dye wastewater treatment. RSC Adv. **14**(32), 23352–23363 (2024). https://doi.org/10.1039/d4ra03570d
47. Gormley, A.J., Webb, M.A.: Machine learning in combinatorial polymer chemistry. Nat. Res. (2021). https://doi.org/10.1038/s41578-021-00282-3
48. Zhao, Y., Mulder, R.J., Houshyar, S., Le, T.C.: A review on the application of molecular descriptors and machine learning in polymer design. R. Soc. Chem. (2023). https://doi.org/10.1039/d3py00395g
49. Tom, G., et al.: Self-driving laboratories for chemistry and materials science. Am. Chem. Soc. (2024). https://doi.org/10.1021/acs.chemrev.4c00055
50. Ge, W., De Silva, R., Fan, Y., Sisson, S.A., Stenzel, M.H.: Machine learning in polymer research. Wiley (2025). https://doi.org/10.1002/adma.202413695
51. Coveney, P.V., Highfield, R.: Artificial intelligence must be made more scientific. Am. Chem. Soc. (2024). https://doi.org/10.1021/acs.jcim.4c01091
52. Bewersdorff, A., Seßler, K., Baur, A., Kasneci, E., Nerdel, C.: Assessing student errors in experimentation using artificial intelligence and large language models: a comparative study with human raters. Comput. Educ.: Artif. Intell. **5** (2023). https://doi.org/10.1016/j.caeai.2023.100177
53. Gaa, I.E., Hohenbergt Ecole, P., Superzeure, X., I'aris, F., Konnt, W.: Phys. Rev.
54. Hollingsworth, S.A., Dror, R.O.: Molecular dynamics simulation for all. Cell Press (2018). https://doi.org/10.1016/j.neuron.2018.08.011
55. Hassija, V., et al.: Interpreting black-box models: a review on explainable artificial intelligence. Springer (2024). https://doi.org/10.1007/s12559-023-10179-8
56. Ferji, K.: Basic concepts and tools of artificial intelligence in polymer science. R. Soc. Chem. (2025). https://doi.org/10.1039/d5py00148j
57. Wang, Y., Lamim Ribeiro, J.M., Tiwary, P.: Machine learning approaches for analyzing and enhancing molecular dynamics simulations. Elsevier Ltd. (2020). https://doi.org/10.1016/j.sbi.2019.12.016
58. Pugliese, R., Regondi, S., Marini, R.: Machine learning-based approach: global trends, research directions, and regulatory standpoints. KeAi Communications Co. (2021). https://doi.org/10.1016/j.dsm.2021.12.002
59. Badini, S., Regondi, S., Pugliese, R.: Unleashing the power of artificial intelligence in materials design. Multidisciplinary Digital Publishing Institute (MDPI) (2023). https://doi.org/10.3390/ma16175927
60. Unke, O.T., et al.: Machine learning force fields. Am. Chem. Soc. (2021). https://doi.org/10.1021/acs.chemrev.0c01111
61. Jain, A., et al.: Commentary: the materials project: A materials genome approach to accelerating materials innovation. American Institute of Physics Inc. (2013). https://doi.org/10.1063/1.4812323
62. Curtarolo, S., et al.: AFLOWLIB.ORG: a distributed materials properties repository from high-throughput ab initio calculations. Comput. Mater. Sci. **58**, 227–235 (2012). https://doi.org/10.1016/j.commatsci.2012.02.002
63. Kirklin, S., et al.: The open quantum materials database (OQMD): assessing the accuracy of DFT formation energies. NPJ Comput. Mater. **1** (2015). https://doi.org/10.1038/npjcompumats.2015.10
64. Otsuka, S., Kuwajima, I., Hosoya, J., Xu, Y., Yamazaki, M.: PoLyInfo: polymer database for polymeric materials design. In: Proceedings—2011 International Conference on Emerging Intelligent Data and Web Technologies, EIDWT 2011, pp. 22–29 (2011). https://doi.org/10.1109/EIDWT.2011.13
65. "60"

66. Lee, D.E., Song, S.-O., Yoon, E.S.: Weighted support vector machine for quality estimation in polymerization processes (2004)
67. Malashin, I., Tynchenko, V., Gantimurov, A., Nelyub, V., Borodulin, A.: Support vector machines in polymer science: a review. Multidisciplinary Digital Publishing Institute (MDPI) (2025). https://doi.org/10.3390/polym17040491
68. Savage, N.: Neural net worth. Commun. ACM **62**(6), 10–12 (2019). https://doi.org/10.1145/3323872
69. Long, T., et al.: Recent progress of artificial intelligence application in polymer materials. (Multidisciplinary Digital Publishing Institute (MDPI)) (2025). https://doi.org/10.3390/polym17121667
70. Breiman, L: Random forests (2001)
71. Tolles, J., Meurer, W.J.: Logistic regression relating patient characteristics to outcomes JAMA guide to statistics and methods (2016). http://jama.jamanetwork.com/
72. Ertefaie, A., McKay, J.R., Oslin, D., Strawderman, R.L.: Robust Q-learning. J. Am. Stat. Assoc. **116**(533), 368–381 (2021). https://doi.org/10.1080/01621459.2020.1753522
73. Orsenigo, C., Vercellis, C.: Kernel ridge regression for out-of-sample mapping in supervised manifold learning. Expert Syst. Appl. **39**(9), 7757–7762 (2012). https://doi.org/10.1016/j.eswa.2012.01.060
74. Jakob, P., Madan, M., Schmid-Schirling, T., Valada, A.: Multi-perspective anomaly detection (2021). https://doi.org/10.3390/s21165311
75. Schulz, E., Speekenbrink, M., Krause, A.: A tutorial on Gaussian process regression: modelling, exploring, and exploiting functions. J. Math. Psychol. **85**, 1–16 (2018). https://doi.org/10.1016/j.jmp.2018.03.001
76. Wang, X., Huang, F., Cheng, Y.: Computational performance optimization of support vector machine based on support vectors. Neurocomputing **211**, 66–71 (2016). https://doi.org/10.1016/j.neucom.2016.04.059
77. del Rio, B.G., Phan, B., Ramprasad, R.: A deep learning framework to emulate density functional theory. NPJ Comput. Mater. **9**(1) (2023). https://doi.org/10.1038/s41524-023-01115-3
78. Yao, K., Herr, J.E., Parkhill, J.: The many-body expansion combined with neural networks. J. Chem. Phys. **146**(1) (2017). https://doi.org/10.1063/1.4973380
79. Gautham, S.M.B., Patra, T.K.: Deep learning potential of mean force between polymer grafted nanoparticles
80. Fang, X., et al.: Universal phase identification of block copolymers from physics-informed machine learning (2024). https://doi.org/10.26434/chemrxiv-2024-9zjt0
81. Li, H., et al.: Tuning the molecular weight distribution from atom transfer radical polymerization using deep reinforcement learning. Mol. Syst. Des. Eng. **3**(3), 496–508 (2018). https://doi.org/10.1039/c7me00131b
82. "132"
83. Webb, M.A., Jackson, N.E., Gil, P.S., De Pablo, J.J.: Targeted sequence design within the coarse-grained polymer genome (2020). http://advances.sciencemag.org/
84. Zhu, C., Zhang, J.: Inferential estimation of polymer melt index using deep belief networks. In: ICAC 2018—2018 24th IEEE International Conference on Automation and Computing: Improving Productivity through Automation and Computing, Institute of Electrical and Electronics Engineers Inc. (2018). https://doi.org/10.23919/IConAC.2018.8749111
85. Sattari, K., Xie, Y., Lin, J.: Data-driven algorithms for inverse design of polymers. R. Soc. Chem. (2021). https://doi.org/10.1039/d1sm00725d
86. Queen, O., et al.: Polymer graph neural networks for multitask property learning. NPJ Comput. Mater. **9**(1) (2023). https://doi.org/10.1038/s41524-023-01034-3
87. Pilania, G., Iverson, C.N., Lookman, T., Marrone, B.L.: Machine-learning-based predictive modeling of glass transition temperatures: a case of polyhydroxyalkanoate homopolymers and copolymers. J. Chem. Inf. Model. **59**(12), 5013–5025 (2019). https://doi.org/10.1021/acs.jcim.9b00807

88. Lee, C.K., et al.: Transfer learning with graph neural networks for optoelectronic properties of conjugated oligomers. J. Chem. Phys. **154**(2) (2021). https://doi.org/10.1063/5.0037863
89. Phan, B.K., Shen, K.H., Gurnani, R., Tran, H., Lively, R., Ramprasad, R.: Gas permeability, diffusivity, and solubility in polymers: simulation-experiment data fusion and multi-task machine learning. NPJ Comput. Mater. **10**(1) (2024). https://doi.org/10.1038/s41524-024-01373-9
90. Upadhya, R., Tamasi, M.J., Di Mare, E., Murthy, N.S., Gormley, A.J.: Data-driven design of protein-like single-chain polymer nanoparticles
91. Khanzadeh, M., Rao, P., Jafari-Marandi, R., Smith, B.K., Tschopp, M.A., Bian, L.: Quantifying geometric accuracy with unsupervised machine learning: using self-organizing map on fused filament fabrication additive manufacturing parts. J. Manuf. Sci. Eng. Trans. ASME **140**(3) (2018). https://doi.org/10.1115/1.4038598
92. Roy, M., Wodo, O.: Data-driven modeling of thermal history in additive manufacturing. Addit. Manuf. **32** (2020). https://doi.org/10.1016/j.addma.2019.101017
93. Lee, X.Y., Saha, S.K., Sarkar, S., Giera, B.: Automated detection of part quality during two-photon lithography via deep learning. Addit. Manuf. **36** (2020). https://doi.org/10.1016/j.addma.2020.101444
94. Lecun, Y., Bengio, Y., Hinton, G.: Deep learning (Nature Publishing Group) (2015). https://doi.org/10.1038/nature14539
95. Li, C., et al.: Rapid Bayesian optimisation for synthesis of short polymer fiber materials. Sci. Rep. **7**(1) (2017). https://doi.org/10.1038/s41598-017-05723-0
96. Takasuka, S., et al.: Bayesian optimization of radical polymerization reactions in a flow synthesis system. Sci. Technol. Adv. Mater.: Methods **4**(1) (2024). https://doi.org/10.1080/27660400.2024.2425178
97. Aizawa, T.: New design method for fabricating multilayer membranes using CO_2-assisted polymer compression process. Molecules **25**(24) (2020). https://doi.org/10.3390/MOLECULES25245786
98. Barnett, J.W., et al.: Designing exceptional gas-separation polymer membranes using machine learning (2020). https://www.science.org
99. Rizkin, B.A., Shkolnik, A.S., Ferraro, N.J., Hartman, R.L.: Combining automated microfluidic experimentation with machine learning for efficient polymerization design. Nat. Mach. Intell. **2**(4), 200–209 (2020). https://doi.org/10.1038/s42256-020-0166-5
100. Chitre, A., et al.: Accelerating formulation design via machine learning: generating a high-throughput shampoo formulations dataset. Sci. Data **11**(1) (2024). https://doi.org/10.1038/s41597-024-03573-w
101. Knox, S.T., Parkinson, S.J., Wilding, C.Y.P., Bourne, R.A., Warren, N.J.: Autonomous polymer synthesis delivered by multi-objective closed-loop optimisation. Polym. Chem. **13**(11), 1576–1585 (2022). https://doi.org/10.1039/d2py00040g
102. Cheng, J., Tan, W., Yuan, Y., Zhao, Z., Cheng, Y.: Research on defect detection method for composite materials based on deep learning networks. Appl. Sci. (Switzerland) **14**(10) (2024) https://doi.org/10.3390/app14104161
103. Gong, Y., Shao, H., Luo, J., Li, Z.: A deep transfer learning model for inclusion defect detection of aeronautics composite materials. Compos. Struct. **252** (2020). https://doi.org/10.1016/j.compstruct.2020.112681
104. Horvitz, E., Mulligan, D.: Data, privacy, and the greater good. Science (1979) **349**(6245), 253–255 (2015). https://doi.org/10.1126/science.aac4520
105. "31"
106. Domingos, P.: A few useful things to know about machine learning (2012). https://doi.org/10.1145/2347736.2347755
107. Wilson, A.N., et al.: PolyID: artificial intelligence for discovering performance-advantaged and sustainable polymers. Macromolecules **56**(21), 8547–8557 (2023). https://doi.org/10.1021/acs.macromol.3c00994
108. López-Rios de Castro, R., Ziolek, R.M., Ulmschneider, M.B., Lorenz, C.D.: Therapeutic peptides are preferentially solubilized in specific microenvironments within PEG-PLGA

polymer nanoparticles. Nano Lett. **24**(6), 2011–2017 (2024). https://doi.org/10.1021/acs.nanolett.3c04558
109. Park, N.H., Zubarev, D.Y., Hedrick, J.L., Kiyek, V., Corbet, C., Lottier, S.: A recommender system for inverse design of polycarbonates and polyesters. Macromolecules **53**(24), 10847–10854 (2020). https://doi.org/10.1021/acs.macromol.0c02127
110. Wheatle, B.K., Fuentes, E.F., Lynd, N.A., Ganesan, V.: Design of polymer blend electrolytes through a machine learning approach. Macromolecules **53**(21), 9449–9459 (2020). https://doi.org/10.1021/acs.macromol.0c01547
111. Bone, J.M., et al.: Hierarchical machine learning for high-fidelity 3D printed biopolymers. ACS Biomater. Sci. Eng. **6**(12), 7021–7031 (2020). https://doi.org/10.1021/acsbiomaterials.0c00755
112. Chen, Z., et al.: Janus bottlebrush compatibilizers. Soft Matter **616** (2024). https://doi.org/10.1039/d3sm01484c
113. Miccio, L.A., Schwartz, G.A.: Mapping chemical structure-glass transition temperature relationship through artificial intelligence. Macromolecules **54**(4), 1811–1817 (2021). https://doi.org/10.1021/acs.macromol.0c02594
114. Lu, Y., Yalcin, D., Pigram, P.J., Blackman, L.D., Boley, M.: Interpretable machine learning models for phase prediction in polymerization-induced self-assembly. J. Chem. Inf. Model. **63**(11), 3288–3306 (2023). https://doi.org/10.1021/acs.jcim.3c00460
115. Jayaraman, A., Olsen, B.: Convergence of artificial intelligence, machine learning, cheminformatics, and polymer science in macromolecules. Am. Chem. Soc. (2024). https://doi.org/10.1021/acs.macromol.4c01704
116. Wu, S., et al.: Machine-learning-assisted discovery of polymers with high thermal conductivity using a molecular design algorithm. NPJ Comput. Mater. **5**(1) (2019). https://doi.org/10.1038/s41524-019-0203-2
117. Aoki, Y., et al.: Multitask machine learning to predict polymer-solvent miscibility using flory-huggins interaction parameters. Macromolecules **56**(14), 5446–5456 (2023). https://doi.org/10.1021/acs.macromol.2c02600
118. Sutliff, B.P., Goyal, S., Martin, T.B., Beaucage, P.A., Audus, D.J., Orski, S.V.: Correlating near-infrared spectra to bulk properties in polyolefins. Macromolecules **57**(5), 2329–2338 (2024). https://doi.org/10.1021/acs.macromol.3c02290
119. Kim, C., Chandrasekaran, A., Huan, T.D., Das, D., Ramprasad, R.: Polymer genome: a data-powered polymer informatics platform for property predictions. Am. Chem. Soc. (2018). https://doi.org/10.1021/acs.jpcc.8b02913
120. Ma, Y., et al.: High-entropy approach versus traditional doping strategy for layered oxide cathodes in alkali-metal-ion batteries: a comparative study. Elsevier B.V. (2025). https://doi.org/10.1016/j.ensm.2025.104295
121. Glielmo, A., Husic, B.E., Rodriguez, A., Clementi, C., Noé, F., Laio, A.: Unsupervised learning methods for molecular simulation data. Am. Chem. Soc. (2021). https://doi.org/10.1021/acs.chemrev.0c01195
122. Botvinick, M., Ritter, S., Wang, J.X., Kurth-Nelson, Z., Blundell, C., Hassabis, D.: Reinforcement learning, fast and slow. Elsevier Ltd. (2019) https://doi.org/10.1016/j.tics.2019.02.006
123. Chanani, P., Rezaei, N., Dormiani, K., Shokatian, M., Ata-abadi, N.S.: Progress and prospect of minicircle as a minimized non-viral DNA vector in gene therapy and regenerative medicine. Mol. Ther. Nucleic Acids 102682 (2025). https://doi.org/10.1016/j.omtn.2025.102682
124. Brereton, R.G., Lloyd, G.R.: Support vector machines for classification and regression. R. Soc. Chem. (2010). https://doi.org/10.1039/b918972f
125. Abdi, H., Williams, L.J.: Principal component analysis (2010). https://doi.org/10.1002/wics.101
126. Likas, A., Vlassis, N., Verbeek, J.J.: The global k-means clustering algorithm (2003). www.elsevier.com/locate/patcog
127. Mahesh, B.: Machine learning algorithms—A review. Int. J. Sci. Res. (IJSR) **9**(1), 381–386 (2020). https://doi.org/10.21275/art20203995

128. Farnham, B., Tokyo, S., Boston, B., Sebastopol, F., Beijing, T.: Hands-on Machine Learning with Scikit-Learn, Keras, and TensorFlow Concepts, Tools, and Techniques to Build Intelligent Systems, 2nd edn.
129. Rosenblatt, F.: The perceptron: a probabilistic model for information storage and organization in the brain **1**
130. Lecun, Y., Bottou, E., Bengio, Y., Haffner, P.: Gradient-based learning applied to document recognition (1998)
131. Hochreiter, S, Urgen Schmidhuber, J.: Long short-term memory
132. Hinton, G.E., Osindero, S.: A fast learning algorithm for deep belief nets Yee-Whye Teh
133. Qu, E.Z., Jimenez, A.M., Kumar, S.K., Zhang, K.: Quantifying nanoparticle assembly states in a polymer matrix through deep learning. Macromolecules **54**(7), 3034–3040 (2021). https://doi.org/10.1021/acs.macromol.0c02483
134. Saxena, A., et al.: A review of clustering techniques and developments. Neurocomputing **267**, 664–681 (2017). https://doi.org/10.1016/j.neucom.2017.06.053
135. Solorio-Fernández, S., Carrasco-Ochoa, J.A., Martínez-Trinidad, J.F.: A review of unsupervised feature selection methods. Artif. Intell. Rev. **53**(2), 907–948 (2020). https://doi.org/10.1007/s10462-019-09682-y
136. Li, H., et al.: Tuning the molecular weight distribution from atom transfer radical polymerization using deep reinforcement learning
137. Pittaway, P.M., et al.: Self-driving laboratory for emulsion polymerization. Chem. Eng. J. **507** (2025). https://doi.org/10.1016/j.cej.2025.160700

Educational and Training Tools for AI in Polymer Science

Megha Bhandari, Neha Kathayat, Netra Pal Sharma, Pankaj Bhatt, Deepak Chandra Melkani, Seeta Dewali, Swetapadma Dash, and Satpal Singh Bisht

Abstract The rapid advancement of artificial intelligence (AI) is transforming polymer science by enabling accelerated materials discovery, predictive modeling, and process optimization. This chapter explores educational and training strategies designed to integrate AI into polymer science curricula, ensuring that future materials scientists are equipped with essential computational skills. It discusses the development and implementation of interactive platforms, massive open online courses (MOOCs), and AI-powered tools that enhance student engagement and facilitate hands-on learning. Emphasis is placed on interdisciplinary teaching approaches that merge polymer chemistry, data science, and machine learning concepts. Furthermore, the chapter highlights innovative training frameworks for researchers, fostering collaboration between academia and industry to bridge the AI skills gap. By presenting pedagogical methodologies and practical case studies, this work aims to prepare the next generation of polymer scientists to effectively leverage AI, driving innovation in materials research and sustainable technological development.

Keywords Artificial intelligence · Polymer science · Educational · Training tools · Computational

1 Integrating AI into Polymer Science Curricula

It has been increasingly evident over the last few years that artificial intelligence (AI) is becoming more and more prevalent in numerous aspects of our lives. The term artificial intelligence (AI) encompasses a wide range of computational methods that

M. Bhandari · N. Kathayat · N. P. Sharma · P. Bhatt · D. C. Melkani · S. Dewali (✉)
Department of Zoology, Kumaun University, Nainital, Uttarakhand, India
e-mail: seemadeoli3@gmail.com

S. Dash
Zoological Survey of India (ZSI), Kolkata, India

S. S. Bisht
Soban Singh Jeena University Almora, Almora, Uttarakhand, India

B. Sharma et al. (eds.), *AI-Driven Polymer Design*, Engineering Materials,
https://doi.org/10.1007/978-981-95-8353-9_11

provide machines the ability to evaluate data, spot patterns, and make predictions that are beyond the capabilities of humans [1, 2]. Machine Learning (ML), a branch of artificial intelligence (AI) that enables computers to learn from data and improve predictions without explicit programming, lies at the heart of this transformation [3–5]. In particular, artificial intelligence (AI) refers to algorithms that can interact with their surroundings in a way that mimics and exceeds human capabilities [6].

AI includes a variety of skills such as problem-solving, natural language comprehension, experience-based learning, and decision-making [7]. As it develops further, it becomes essential in a number of industries, such as healthcare, banking, and education [8]. We can utilize data to create predictive models for intricate events and use these models in generative situations to generate ideas or recommendations because of the sophisticated algorithms and modelling techniques clustered under the broad umbrella of "AI" [9]. Image-processing models can categorize medical images to determine health or disease states, while generative models can generate images from text prompts [10]. AI models can also predict the properties of small molecules and generative models can suggest synthetic routes to most efficiently make those molecules by quantifying the likelihood of success of various reaction pathways [11].

A widely used approach in the branch of artificial intelligence called machine learning (ML) is to directly or iteratively provide solutions depending on a set of parameters [12]. One or more scores are used to evaluate the results, and the algorithm either aims to maximize these scores or predicts them using combinations of parameters. An ML algorithm that iteratively refines the synthesis process based on feedback to optimize certain or different properties of a polymer is a pertinent example [13]. Artificial intelligence (AI) systems and Machine learning (ML) models and have enormous potential to speed up the development of polymeric materials by drastically lowering the time and labor costs required to find more optimal material platforms [14].

AI involves more than just computations; it also involves learning, adapting, and making judgments on its own using data [15]. Beyond simple automation, artificial intelligence (AI) in polymer research includes (Fig. 1) data-driven material discovery, predictive modeling, and self-optimizing synthesis [5, 16]. In contrast to Molecular Dyanamics simulations, Bhattacharya and Patra [17] showed that AI could predict polymer phase transitions, including the coil-to-globule transition, with high accuracy and at a substantially lower computing cost. Instead of taking the place of these conventional techniques, AI is a potent addition that improves our capacity to effectively investigate intricate polymer systems [18].

The enormous potential advantages of machine learning (ML) systems for polymer science have led to a great deal of model development for a wide range of applications. These include basic inverse design of materials with specified qualities, as well as specialized applications such as cloud point engineering, gas separation, thermal conductivity, mechanical toughness, MRI contrast agents, and polymer electrolytes [19, 20]. There are many barriers to the regular use of ML models in polymer research operations like limited data availability, accuracy, diversity, volume

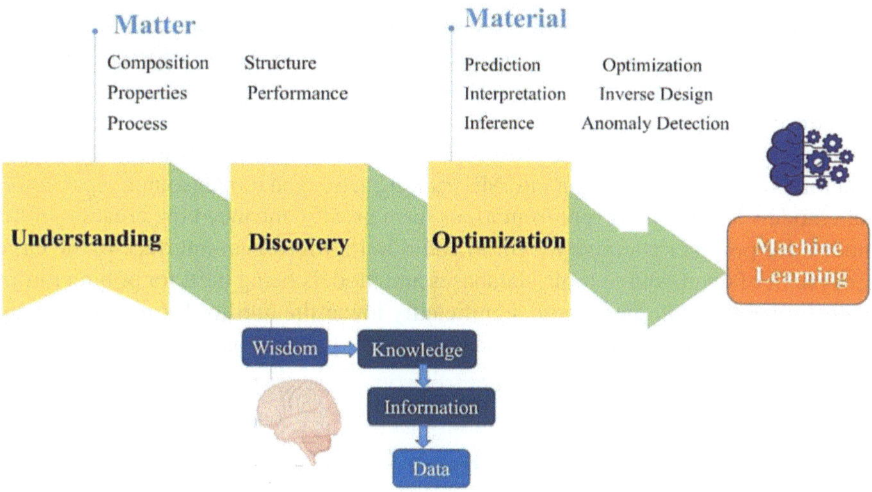

Fig. 1 Machine learning paradigm for polymer science: from matter understanding to material optimization

and relevance for the creation of strong machine learning models have been caused in part by the lack of a developed open source polymer data ecosystem [21].

As their individual data volumes rises, recent developments in open-source data repositories for polymeric materials, like the Polymer Genome or CRIPT, will significantly help to mitigate this problem [14]. Machine learning is also used for estimating polymer properties like size [22]. When ML is utilized to forecast the collision cross-section of polymers, as in ion mobility spectrometry, these predictions can be helpful in identification procedures [23]. The efficiency and precision of the identification process are improved by these predictions, which help to reduce the search window for possible candidates [24].

Polymer research has advanced for about a century and a developing theoretical framework for comprehension, design, preparation, and application has been established [20, 25]. We must reexamine those appealing models for polymer science in the association of molecular traits with performance and properties at the macroscale due to the urgent need of quantitative data in the exploding era of artificial intelligence [26].

Over the past 20 years, there have been significant advancements in the prediction of polymer properties. Initially, inference was used, followed by additive group contributions [27], nonlinear and synergistic contributions in structure–property correlations and finally, machine learning techniques were used to discover new polymers [28, 29].

In order to improve measurable qualities and maybe other relevant features, ML is very effective at optimizing synthesis conditions [30, 31]. The development of polymer-based biomaterials polymeric long-acting injectables and orodispersible films for drug administration are noteworthy instances of these uses [32–34], determining the adjustable variables and the qualities of interest (polymer properties) is a prerequisite for carrying out these optimizations [35]. The monomer conversion, where ideally all of the monomers are converted, may be a property of interest for polymer synthesis. This was accomplished by connecting nuclear magnetic resonance (NMR) to a flow reactor [36].

There is a need for software tools that offer flexibility in experimental data representations and their translation into ML training sets, even if opensource repositories, data models and polymer representations have greatly enhanced the creation of ML models for polymer chemistry [16]. In addition to offering an interface to the larger ecosystem of open-source tools, databases and models being built for polymer informatics, such technologies would significantly lower the hurdles for research groups to start using their own historical datasets in ML applications [5].

2 Importance of AI in Polymer Science

As AI is gaining importance its role may be misinterpreted particularly for new users in domains like polymer science where it could be mistaken for traditional modelling methods or industrial automation [37]. It could be difficult for researchers who are used to more conventional computational techniques to distinguish AI from explicitly coded simulations or automated control systems [38]. This could cause ambiguity regarding the true meaning of artificial intelligence and how it varies from other digital tools. Because of this, many people would believe that any computational system—from factory sensors to molecular simulations—would be regarded as artificial intelligence (AI). This misconception could make it difficult to distinguish between data-driven intelligence and conventional computing [39, 40]. AI is all about learning, adapting, and making judgments on its own based on data, not just about doing calculations [41, 42]. In polymer research AI is beyond simple automation, it includes data-driven material discovery, predictive modelling, and self-optimizing synthesis.

AI explicitly identifies patterns in data, thus, presents a completely new paradigm. This makes it possible to make precise predictions even in cases when the underlying physics is not completely known [43] Many AI methods, in contrast to conventional models, operate as "black boxes," generating predictions based on intricate statistical correlations rather than clear-cut physical rules [44]. Because AI-generated results may be hard to explain using physical principles, this lack of interpretability might cause resentment among the scientific community. AI is an effective complement to these conventional techniques, not a replacement for them, improving our capacity to effectively investigate intricate polymer systems [45]. AI is not a panacea, though; the caliber and variety of the data it is trained on directly affects how accurate the data is [46]. Inaccurate forecasts from poorly managed datasets can result in inaccurate results, ultimately reducing AI's dependability in real-world scenarios [47]. The synthesis and characterisation of polymers are beginning to be significantly impacted by the fast growing applications of machine learning (ML) and artificial intelligence (AI) [11, 48].

3 Four Stages for the Application of Machine Learning Approach in Polymer Science

There are typically four steps involved in integrating ML techniques in polymer science: data collection, feature and data engineering, model creation, evaluation, and application [23].

Acquiring a high-quality dataset from a variety of sources, such as literature, expert-annotated databases, handbooks, standards and product datasheets, is essential for applying machine learning techniques in polymer science [49, 50]. Experimental measurements, theoretical computations, physical modeling and simulation, intrinsic qualities and expert labels in dictionary forms are some possible sources of these data [20]. A lot of work has been done to make it easier to gather scattered data from various sources in order to create high quality datasets, like large language models based on entity recognition [51], and to extract text-based data, like polymer names, compositions, and properties, in a steady manner, therefore it is possible to consistently extract text-based information, including polymer names, compositions, and features [52].

The process of data engineering involves gathering fragmented raw data from multimodality sources, annotating the data in a single pipeline (Fig. 2), fusing dictionary data without taking into account their original intent and then assembling the data into a structured dataset for knowledge and wisdom illustration, statistical analysis, and machine learning [53]. The annotation of large, representative, consistent, and accurate datasets for public access has advanced significantly. Examples include the Protein Data Bank [54], PubChem [55], ChemSpider [56], Reaxys and QM9 [57, 58], and others. In particular, polymers have PolyInfo [59], Polymer Genome [28], and Polymer Property Database [28]. The goal of feature engineering is to model targets by assembling a wealth of physically well-defined and informative features [60]. The targets can be either the formulation or optimum processing in inverse design from a known combination of needed attributes, or they can be 3D structures, properties, and performance metrics in forward prediction with known composition, process, and structural information [20]. Data and feature engineering are usually carried out to enhance data quality.

In polymer science, correlations can be evaluated using linear, monotonous, periodic, reciprocal, or other implicit correlations, and can involve several correlated components [61]. According to the No Free Lunch (NFL) theorem, there is currently no method or algorithm that can produce a universally robust model for the quantitative CPSPPr for any polymers [62]. This is because the existence of a nonlinear, synergistic, and far from equilibrium state presents significant challenges for conventional approaches to present polymers [63].

ML models must undergo rigorous review prior to use since they are continually challenged by the bias variance dilemma. The most important factors for evaluating an ML model are its interpretability, robustness, accuracy, and precision [20].

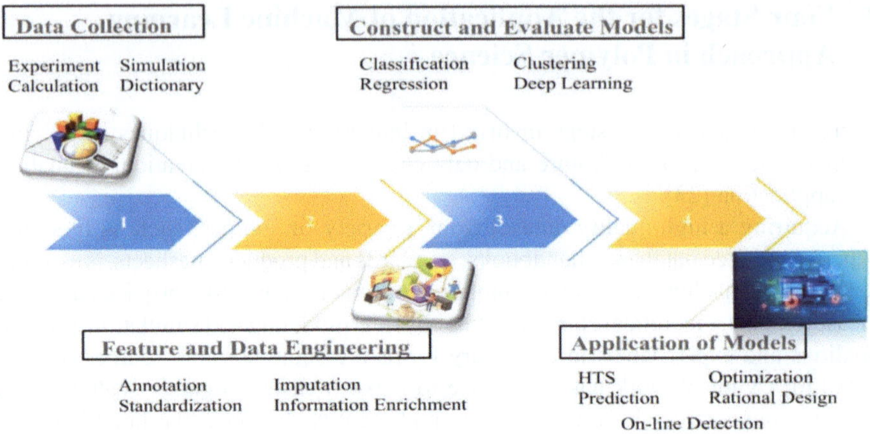

Fig. 2 Pipeline for applying machine learning approaches in polymer research

4 Types of Machine Learning

A fundamental branch of artificial intelligence called machine learning (ML) aims to create models that can learn from data and make judgments or predictions without the need for explicit programming [64]. ML is being utilized progressively in polymer research to guide material discovery, optimize synthesis and predict characteristics [65]. The three primary types of machine learning (ML) are reinforcement learning (RL), unsupervised learning (UL), and supervised learning (SL). These methods differ in the kinds of problems they address, the amount of human oversight needed and the way they handle data [5, 66, 67].

4.1 Supervised Learning

Models learn from labeled datasets in supervised learning (SL), while each input has a known output [68]. This method is comparable to conventional classroom instruction, in which an instructor guides students' learning via examples and corrections. An SL model, in contrast to human learning, continuously assesses its performance and iteratively modifies its parameters until it achieves a high-performance threshold, guaranteeing accurate predictions [69].

The two main task for which SL is utilized for are regression and classification [70]. For Classification task, the model can distinguish between biodegradable and non-biodegradable polymers and predict categorical results [71]. The biodegradability of novel polymers can be accurately predicted by an SL model by learning patterns from a dataset that includes chemical structures and degradation parameters [72].

In polymer research, SL has been to leverage extensive experimental datasets to address challenging material issues. One application is predicting the compatibility of polymers with solvents [73], Complex data analysis tasks, particularly those involving microscope image processing, have been automated using SL [74]. A significant challenge in polymer nanocomposite research is the accurate localization and characterization of nanoparticles within polymer matrices, which is typically performed manually or through time-intensive image analysis tools. To address this issue, Qu et al. [75] developed a deep learning-based method to detect and quantify nanoparticles in transmission electron microscopy (TEM) images.

4.2 *Unsupervised Learning*

Unsupervised learning (UL) is a potent method that detects patterns in unlabeled data—that is, data without any predetermined outputs [76]. UL models examine data on their own to find hidden structures, clusters, or correlations [77], in contrast to supervised learning, which depends on explicit input–output pairs. To put it another way, it's similar to a student studying texts on their own to find recurring themes without assistance from a teacher. UL is especially useful for deciphering complicated polymer datasets where experimental labels could be hard to find or describe [78]. UL is especially helpful for dimensionality reduction, which streamlines high-dimensional polymer datasets while maintaining crucial information, and clustering, polymers with similar structural or chemical traits [79, 80]. To understand more about the polymer conformations inside pluronic core–shell micelles, [81] employed unsupervised machine learning techniques.

Polymer research has effectively used UL approaches to gain valuable insights from intricate datasets [82]. Ziolek et al. [81] examined the nanoscale structure of micelles made of four-arm and linear block copolymers using UL techniques. They found groups of micelle structures with comparable corona configurations by clustering molecular conformations, and dimensionality reduction made the intricate structural changes easier to understand. Deeper understanding of self-assembly mechanisms—which are essential for the creation of biomaterials and medication delivery—was made possible by their methodology.

4.3 *Reinforcement Learning*

A unique subset of machine learning referred to as reinforcement learning (RL) where models learns by interactions with the environment and gets rewards for taking optimal actions [83]. Through trial and error, RL algorithms find the best solutions, which makes them especially well-suited for tasks that require for sequential decision-making [84]. An effective comparison is when a young child discovers that fire is deadly only after coming into contact with it; this knowledge is acquired through firsthand experience rather than previous education [85]. RL is far more complex than supervised and unsupervised learning because it incorporates exploration–exploitation trade-offs, long-term reward optimization, and sequential decision making [86]. RL constantly modifies strategies based on ongoing input, in contrast to models that learn from static datasets. This necessitates sophisticated algorithms and substantial processing resources. These characteristics add to RL's increased mathematical and practical complexity while simultaneously making it an effective tool for autonomous experimental control and polymerization process optimization [87–90].

RL involves high-fidelity simulation environments or substantial real-time experimentation, both of which need a lot of resources [91]. Furthermore, it can be challenging to define appropriate action spaces and reward functions for polymer systems. Because of this, RL might work better on challenges that are more specialized (such improving a particular polymerization procedure) as opposed to exploratory tasks that are more general.

Whether for predicting properties accurately, revealing hidden patterns, or facilitating self-learning experimental processes, each machine learning technique—supervised, unsupervised, and reinforcement learning—offers unique possibilities. The learning process, computational complexity, and application range of these approaches vary. Many AI tools and platforms are available to help data administration, analysis, and modelling in order to make it easier to integrate machine learning (ML) into polymer research [5].

The use of cheminformatics tools to process molecular representations is a major application of machine learning in polymer research [92]. Chemical structures are transformed by RDKit into machine-readable representations, like molecular fingerprints or SMILES strings, which are then used as inputs for machine learning models [93]. This capability is extended to stochastic polymers by Big SMILES, which makes it possible to depict structural differences in polymer chains. In the meantime, Polymer Genome provides pretrained models for predicting polymer properties, which makes it easier to quickly screen polymer candidates using molecular descriptors [94].

5 Interdisciplinary Approach for Fulfilling Educational Gap in Polymer Science

Many researchers who are accustomed to empirical methodologies or conventional computational procedures will find the introduction of AI in polymer science to be a fundamental change [11]. While short-term partnerships between AI specialists and polymer scientists aid in bridging this divide, including AI and ML instruction into polymer science curricula is the long-term solution [21]. It is frequently more practicable to teach AI and ML to polymer researchers rather than computer scientists polymer chemistry and engineering because of the specialized nature of

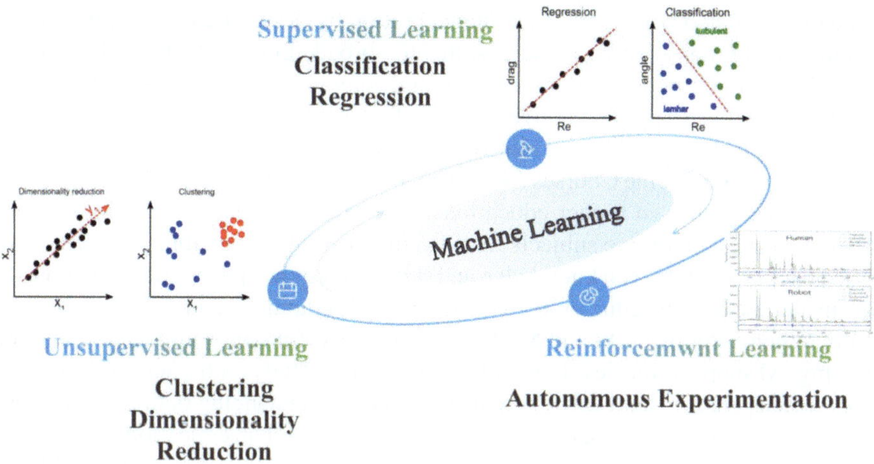

Fig. 3 Machine learning approaches in polymer science: supervised, unsupervised, and reinforcement learning

polymer science and its experimental complexities [95]. There is still a lack of structured AI instruction in polymer science courses, despite AI's increasing influence on materials research. The number of researchers who can advance ML-driven polymer research is limited by the lack of specialized training that combines polymer science and data-driven techniques in master's degrees. In addition to impeding academic advancement, this educational disparity has an impact on the polymer sector, where there is a growing need for interdisciplinary knowledge [5].

In Fig. 3 AI and polymer science together open up new possibilities, such as autonomous research workflows and predictive modelling [37]. In the future, self-optimization of polymerization processes and the deciphering of intricate structure–property correlations are anticipated to be more heavily influenced by reinforcement learning and unsupervised learning [96]. AI will drastically change polymer research, moving away from empirical iteration and toward data-driven hypothesis development, rather than just accelerating the methods [5].

5.1 Interactive Platforms, MOOCs, and AI Tools for Student and Researcher Training

Artificial Intelligence (AI) and Massive Open Online Courses (MOOCs) intersect powerfully in the field of education. With AI, student data can be analyzed to personalize learning pathways, ensuring that course material matches each learner's pace and needs [137]. This adaptability enhances the overall effectiveness of MOOCs. In large-scale online courses, where traditional grading becomes unmanageable, AI can automate assessments and provide timely, individualized feedback. Beyond evaluation, AI also contributes to content creation by designing quizzes, interactive modules, and other learning resources, thereby broadening the scope and quality of MOOCs. By tracking learner engagement and completion patterns, AI helps educators refine course design and delivery. It also fosters inclusivity tools such as speech recognition, text-to-speech, and real-time translation [138, 139].

Massive Open Online Courses (MOOCs), introduced in the early 2010s, represent a revolutionary shift in higher education. Built to be globally accessible, MOOCs provide courses on diverse subjects from leading institutions, enabling learners from varied backgrounds to gain knowledge and skills at their own pace. Their core mission lies in democratizing education by removing geographical and financial barriers [140]. When combined, AI and MOOCs create powerful opportunities to transform learning. AI supports the development of personalized study paths, automates routine tasks such as grading, and delivers real-time feedback enhancing both efficiency and engagement. Together, these technologies pave the way for an educational ecosystem that is adaptive, inclusive, and responsive to the evolving needs of learners worldwide [141, 142].

One of the most prominent themes is the role of AI in personalizing learning, where algorithms analyze student performance and behavior to recommend adaptive learning paths, significantly improving engagement and completion rates [143]. AI is also widely recognized for its ability to automate assessment and feedback, with natural language processing and machine learning enabling immediate evaluation of assignments, thereby reducing instructor workload and enhancing the student experience. Another key area of research involves AI-driven learning analytics, which processes large volumes of learner data to identify behavioral patterns, predict outcomes, and inform course design for better retention. Beyond evaluation, AI contributes to content creation by generating quizzes, interactive exercises, and high-quality learning materials tailored to diverse learner needs. Furthermore, AI-powered chatbots and virtual assistants play an increasingly important role in MOOCs, providing instant support, guiding learners, and fostering engagement, which positively impacts motivation and retention [144].

A variety of tools and platforms are being developed to integrate Artificial Intelligence (AI) into Massive Open Online Courses (MOOCs), with the goal of improving learning experiences and streamlining course management. Adaptive learning platforms such as *Smart Sparrow* and *Knewton* are widely used to personalize content delivery, tailoring resources and activities to the specific needs and performance of individual learners. in Fig. 4 AI-powered assessment tools further support instructors and students by automating evaluation processes—*Gradescope*, for instance, applies AI to grading assignments and exams efficiently, while *Turnitin* provides plagiarism detection and writing feedback to uphold academic integrity [145]. In addition, conversational AI tools such as chatbots and virtual assistants are increasingly embedded in online learning environments, *Moodle Chatbot* offers immediate responses to student queries within the Moodle platform, while *ChatGPT* can serve as a tutor, content generator, or interactive guide [146]. Learning analytics platforms also play a vital role, with systems like *Edvance360* and *Brightspace* using AI to monitor engagement, analyze performance trends, and deliver actionable insights to educators. Collectively, these tools demonstrate how AI is not only enhancing personalization and interactivity in MOOCs but also supporting more efficient, data-driven educational practices.

A wide range of digital tools supports the integration of Artificial Intelligence (AI) with Massive Open Online Courses (MOOCs), enabling richer, more interactive learning environments. For content creation and curation, platforms such as *Canva* provide AI-assisted design features for visually engaging course materials, while *Loom* allows rapid production of video content with AI-based editing enhancements. Core MOOC providers also play a central role: *Coursera* offers diverse courses with assessments and peer interactions, *edX* delivers university-led academic programs with certification options, *Udacity* specializes in technology-driven subjects like programming and data science, and *FutureLearn* emphasizes collaborative, discussion-based learning [147]. In terms of collaboration, tools such as *Slack*, *Microsoft Teams*, and *Google Workspace* (including Google Docs and Drive) facilitate communication, file sharing, and collective content development among learners and instructors [148]. Finally, gamification platforms like *Kahoot!* and *Quizlet* enhance engagement by introducing interactive quizzes, flashcards, and game-based learning experiences. Together, these tools highlight how the thoughtful integration of AI-driven design, collaborative platforms, and gamification strategies can make MOOCs more personalized, engaging, and effective for diverse learner groups.

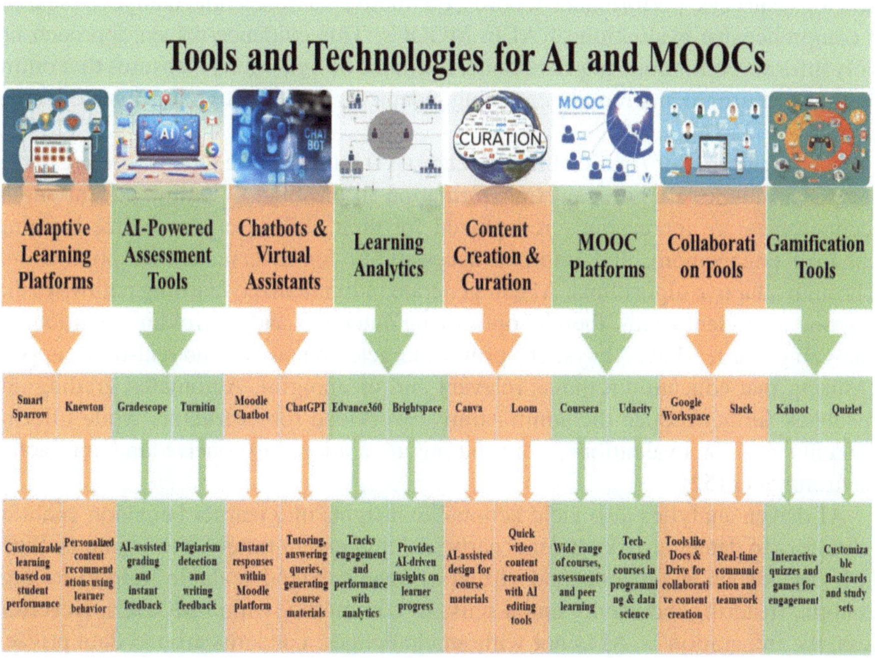

Fig. 4 Interactive platforms, MOOCs, and AI tools in education and research

Data collection is a critical foundation for evaluating how Artificial Intelligence (AI) can improve the effectiveness of Massive Open Online Courses (MOOCs). One of the most widely used methods involves gathering user interaction data through Learning Management Systems (LMS), which track activities such as time spent on videos, navigation patterns, clicks on resources, and assignment submissions [149]. Server log files further provide granular insights into learner engagement and usage trends. Surveys and questionnaires also play a significant role, with pre-course surveys capturing demographic information and prior knowledge, and post-course surveys measuring satisfaction, usability of AI features, and perceived effectiveness.

Incorporating Likert-scale items allows for quantitative analysis of trends and correlations. Performance metrics obtained from quizzes, assignments, and final grades offer another layer of data, capturing completion rates, scores, and time taken to finish tasks. To complement quantitative measures, qualitative methods such as interviews, focus groups, and open-ended survey questions are employed to explore learner and instructor experiences in depth [150]. Additionally, learning analytics techniques analyze clickstream data and engagement metrics, such as forum participation and peer collaboration, to reveal patterns of interaction and dropout points. Experimental designs, including A/B testing, are increasingly used to compare different AI-driven features or course structures, with data collected on engagement, retention, and satisfaction. External sources, such as social media discussions and publicly available educational datasets, further enrich the understanding of learner behavior and attitudes. Taken together, these mixed methods—spanning user interaction logs, surveys, performance records, qualitative feedback, and analytics—enable a comprehensive evaluation of AI in MOOCs. This evidence-driven approach not only informs continuous improvements in course design but also ensures that online learning environments remain engaging, adaptive, and effective for diverse learner populations [151].

The integration of Artificial Intelligence (AI) into Massive Open Online Courses (MOOCs) marks a significant transformation in the field of online education. AI technologies are increasingly recognized for their ability to personalize learning, enhance engagement, and improve accessibility for diverse learner populations. Through adaptive algorithms, AI can generate individualized learning pathways that respond to student needs, thereby optimizing outcomes and improving retention. By analyzing learner data, AI systems provide targeted resources and real-time support, ensuring that education remains relevant and meaningful. Automated grading and feedback further reduce the administrative workload for instructors while offering students prompt evaluations, contributing to a more responsive and interactive environment [152].

AI-driven analytics also yield actionable insights into learner behavior, enabling educators to detect at-risk students early and implement timely interventions. Similarly, tools such as chatbots and virtual assistants extend scalable support to large cohorts, fostering inclusivity and a sense of community within MOOCs. Nevertheless, the integration of AI is not without challenges: concerns around data privacy, algorithmic bias, and the risk of depersonalizing education underscore the need for responsible implementation. Ensuring fairness, transparency, and equitable access to AI tools remains central to their effective use in education [153].

Looking ahead, the synergy between AI and MOOCs offers considerable promise for reshaping digital learning ecosystems. By harnessing these innovations, educators can design more inclusive, engaging, and effective learning experiences that empower students worldwide. To fully realize this potential, sustained research, interdisciplinary collaboration, and ethical vigilance will be essential in guiding the future of AI in education.

5.2 *Preparing the Next Generation of Materials Scientists with AI Skills*

Historically, advances in characterisation, synthesis, and computation technologies have coincided with the development of Material science, Artificial intelligence (AI) and machine learning (ML) are also included [97]. These days, more complex AI models that can tackle more challenging issues in a range of domains are being observed. It is undeniable from the standpoint of material science that artificial intelligence and machine learning provide a powerful toolkit that might significantly speed up research in fields like the creation and discovery of novel functional materials [98]. Artificial Intelligence (AI) encompasses a wide range of concepts, from comparatively basic algorithms to complex universal function approximators [99], which, when combined with robotics, enable autonomous interaction with the physical environment. In essence, regression and classification models are shortcuts that allow one to reasonably predict experimental results [70] or material attributes without synthesizing the material and carrying out the trials [100]. These models can be trained to identify associations between the molecular structure and characteristics like solubility [101], toxicity [102], or antibacterial activity [103], or material composition and solar cell efficiency [104, 105]. After training, these models can be used to virtually screen novel compounds and materials to find intriguing candidates for more in-depth research, significantly lowering the number of experiments required [97].

The primary goal of material science is to investigate the relationships among a material's structure, qualities, process, and use. The advancement of human society will be significantly aided by the discovery of new materials. The area of materials science has developed over several centuries, and a significant amount of data has been gathered [106]. Machine learning (ML) and artificial intelligence (AI) techniques have been hailed as the next scientific paradigm in materials discovery and optimization in recent publications, reviews, symposia, and seminars [18, 107]. Applications of materials science are expanding, encompassing knowledge extraction, experiment selection, and data analysis [108, 109]. The prevalence of machine learning (ML) systems in our daily lives, the free software, and the proven achievements in materials discovery and real-time data collection spurred by the Materials Genome Initiative (MGI) are some of the many factors contributing to this trend [110,

111]. ML has recently been successfully used to materials science, as evidenced by the high-throughput experimental (HTE, or "combinatorial") community. Analysis of hundreds to thousands of high-quality measurements correlated in composition, processing, and microstructure is severely impeded by parallel material production and quick characterization [112, 113]. As the HTE community grew, innovative and novel approaches to property measurement, data visualization, and interpretation were developed. Decision-making and experimental planning are becoming emerging obstacles in the materials research process as algorithms for automating these processes are advance. This indicates that developments in artificial intelligence (AI) for materials research are crucial for boosting material innovation, such as through the development of autonomous experimental systems [114, 115].

Building strong scientific AI systems that can do more than just generate leads—that is, guide people in the correct direction—and offer comprehensive functionality that facilitates scientific discovery will be necessary to go from expectations to practice for AI [116]. There are two ways to reduce this disparity: 1. Creating scientific AI systems that integrate machine learning approaches with physical mechanisms 2. Creative uses of AI systems to produce scientific knowledge directly. There is a need for algorithm development by multidisciplinary researchers of materials science and engineering (MSE) and ML researchers for achieving these two goals. This advancement will be aided by distributed automated laboratory systems, which will provide a platform for highly effective interdisciplinary collaboration by disseminating access to cutting-edge experimental materials research [117]. Utilizing current machine learning techniques effectively has opened up new possibilities to speed up scientific research. Materials scientists may be able to get important insights into why heuristic models and physical theories fail by using algorithmically driven testing to find counterexamples [118].

As a relatively new technology, sophisticated generative artificial intelligence is still being developed, and we are continually learning how it may advance, speed up, and enhance our research. The speed at which these technologies are developing now creates even more uncertainty, making it difficult to predict the full scope of their potential developments [119]. One thing is certain, though: these systems have the capacity to develop into invaluable and revolutionary research instruments. These AI systems could improve scientific writing and public dissemination, speed up the process of finding patterns and trends in scientific literature, help with coding for data analysis and visualization, and—perhaps most importantly—increase our capacity to produce research hypotheses [120].

AI systems have the potential to improve scientific writing and public dissemination, speed up the process of spotting patterns and trends in scientific literature, help with data analysis and visualization coding, and—perhaps most importantly—increase our capacity to produce research hypotheses [121]. It would be quite prudent for researchers who are seeking to remain competitive in the long run to be updated on these technology developments and try out different ways they may be applied to enhance, expand, and supplement the research process. Researchers won't lose their work to an AI agent at this level of intelligence, but they might lose it to researchers who have worked out how to use AI systems to increase the caliber and output of their own study [98]. Even though general artificial intelligence may not be impending, these technologies offer priceless tools that can greatly speed up efforts in material science, such as creating and discovering new functional materials targeted at tackling urgent global concerns [122].

The type of the data gathered and the specific characteristics that are significant are the major factors that set machine learning applications in materials science apart from other fields [123, 124]. In the field of materials science, it is essential to distinguish between theoretical data, experimental data that is accessible to the public, and experimental data that is produced internally [125]. The materials sector often has comparatively small datasets, with a few significant exceptions. The NOMAD, Materials Project, Aflow, and other DFT databases, which may contain information for a few hundred thousand compounds, are among the exceptions [126–128]. These DFT databases are intriguing due to their size as well as the fact that they include information on the crystal structure of the materials [129].

These technologies offer vital tools that can greatly expedite efforts in material science, such as the development and discovery of new functional materials aimed at tackling pressing global concerns, even though general artificial intelligence may not be forthcoming [130]. This calls for a change in the skill sets that researchers need to possess. Data science, programming, a thorough comprehension of the theoretical underpinnings of the chosen subject, and a clear awareness of our goals and the significance of our discoveries are among the talents that we believe will become increasingly valuable [98].

The workforce must be developed immediately in order to make sure that AI methods are incorporated into the materials science workflow with the proper degree of scientific rigor. This can be done by following measures: Teaching the workforce of the future to understand AI methods and how they relate to materials science, Giving present employees more skills so they can effectively mentor the workforce of the future, Adopting a culture that values open data [117].

5.3 Educating the Next-Generation Workforce

For the next generation of workers to be competitive, materials science curricula urgently need to be restructured [131]. This reorganization must consider both direct application to materials research and the degree of transferability of skills required throughout the entire data landscape. There aren't many obligatory courses in programming and statistical approaches in traditional materials science curriculum. The materials science community's adoption of ML approaches is severely hampered lack of fundamental understanding and abilities of graduates [132]. There is critical need for the development of open data/code repositories such as GaTech [133] and nanoHUB [134] that can augment the curriculum of undergraduates.

5.4 *Expanding Skills Within the Current Workforce*

Giving insightful comments on the applicability of models and their results is critically needed at the graduate and post-graduate levels [135]. Materials scientists in their mid- to late-career stages may feel unprepared to guide researchers using machine learning techniques [117]. This can cause people to feel unprepared to apply informed skepticism or to blindly trust (or reject) results from ML procedures.

5.5 *Adopting an Open Data Culture*

Data and metadata are carefully collected and stored in standardized open repositories that are also open to low-capacity research institutes under an open data culture [136]. Community colleges and low-capacity research institutions will be able to join the materials research community and significantly increase diversity by embracing the open data paradigm [117].

6 Conclusion

The integration of artificial intelligence into polymer science education is no longer optional but essential. Embedding AI concepts within curricula provides students with the foundational literacy to engage meaningfully with data-driven research. Complementing formal instruction, interactive platforms, MOOCs, and specialized AI tools expand access to practical training, fostering both independent learning and collaborative exploration. Together, these approaches prepare the next generation of materials scientists to not only apply AI in solving complex problems but also to innovate at the interface of computation and materials research. By aligning pedagogy with technological advances, the field can cultivate researchers who are equipped to lead in an era where data and discovery are inseparably linked.

References

1. Zhang, C., Lu, Y.: Study on artificial intelligence: the state of the art and future prospects. J. Ind. Inf. Integr. **23**, 100224 (2021)
2. Xu, Y., Liu, X., Cao, X., Huang, C., Liu, E., Qian, S., Liu, X., Wu, Y., Dong, F., Qiu, C.W., Qiu, J.: Artificial intelligence: a powerful paradigm for scientific research. Innovation **2**(4) (2021)
3. Mehrotra, D.: Basics of Artificial Intelligence & Machine Learning. Notion Press (2019)
4. Panat, S., Kumar, R.: Introduction to artificial intelligence & ML. In: A Guide to Applied Machine Learning for Biologists, pp. 127–146. Springer International Publishing, Cham (2023)

5. Ferji, K.: Basic concepts and tools of artificial intelligence in polymer science. Polym. Chem. **16**(21), 2457–2470 (2025)
6. Russell, S.J., Norvig, P.: Artificial Intelligence: a Modern Approach. Pearson (2016)
7. Deckker, D., Sumanasekara, S.: AI in vocational and technical education: revolutionizing skill-based learning. EPRA Int. J. Multidiscip. Res. (IJMR) **11**(3), 9 (2025). https://doi.org/10.36713/epra20462
8. Yadav, U., Shrawankar, U.: Artificial intelligence across industries: a comprehensive review with a focus on education. In: AI Applications and Strategies in Teacher Education, pp. 275–320 (2025)
9. Singh, J., Singh, D.: A comprehensive review of clustering techniques in artificial intelligence for knowledge discovery: taxonomy, challenges, applications and future prospects. Adv. Eng. Inform. **62**, 102799 (2024)
10. Liu, V., Chilton, L.B.: Design guidelines for prompt engineering text-to-image generative models. In: Proceedings of the 2022 CHI Conference on Human Factors in Computing Systems, pp. 1–23 (2022)
11. Jayaraman, A., Olsen, B.: Convergence of artificial intelligence, machine learning, chem-informatics, and polymer science in macromolecules. Macromolecules **57**(16), 7685–7688 (2024)
12. Sun, S., Cao, Z., Zhu, H., Zhao, J.: A survey of optimization methods from a machine learning perspective. IEEE Trans. Cybern. **50**(8), 3668–3681 (2019)
13. Queen, O., McCarver, G.A., Thatigotla, S., Abolins, B.P., Brown, C.L., Maroulas, V., Vogiatzis, K.D.: Polymer graph neural networks for multitask property learning. NPJ Comput. Mater. **9**(1), 90 (2023)
14. Park, N.H., Manica, M., Born, J., Hedrick, J.L., Erdmann, T., Zubarev, D.Y., Adell-Mill, N., Arrechea, P.L.: Artificial intelligence driven design of catalysts and materials for ring opening polymerization using a domain-specific language. Nat. Commun. **14**(1), 3686 (2023)
15. Tien, J.M.: Internet of things, real-time decision making, and artificial intelligence. Ann. Data Sci. **4**, 149–178 (2017)
16. Zhao, Y., Chen, G., Liu, J.: Polymer data challenges in the AI era: bridging gaps for next-generation energy materials (2025). arXiv:2505.13494
17. Bhattacharya, D., Patra, T.K.: DPOLY: deep learning of polymer phases and phase transition. Macromolecules **54**(7), 3065–3074 (2021)
18. Wang, Y., Ribeiro, J.M.L., Tiwary, P.: Machine learning approaches for analyzing and enhancing molecular dynamics simulations. Curr. Opin. Struct. Biol. **61**, 139–145 (2020)
19. Bin Abu Sofian, A.D.A., Sun, X., Gupta, V.K., Berenjian, A., Xia, A., Ma, Z., Show, P.L.:. Advances, synergy, and perspectives of machine learning and biobased polymers for energy, fuels, and biochemicals for a sustainable future. Energy Fuels **38**(3), 1593–1617 (2024)
20. Xie, C., Qiu, H., Liu, L., You, Y., Li, H., Li, Y., Sun, Z., Lin, J., An, L.: Machine learning approaches in polymer science: progress and fundamental for a New Paradigm. SmartMat **6**(1), e1320 (2025)
21. Martin, T.B., Audus, D.J.: Emerging trends in machine learning: a polymer perspective. ACS Polymers Au **3**(3), 239–258 (2023)
22. Malashin, I.P., Tynchenko, V.S., Nelyub, V.A., Borodulin, A.S., Gantimurov, A.P.: Estimation and prediction of the polymers' physical characteristics using the machine learning models. Polymers **16**(1), 115 (2023)
23. Li, X., Wang, H., Jiang, M., Ding, M., Xu, X., Xu, B., Zou, Y., Yu, Y., Yang, W.: Collision cross section prediction based on machine learning. Molecules **28**(10), 4050 (2023)
24. Foster, M., Rainey, M., Watson, C., Dodds, J.N., Kirkwood, K.I., Fernández, F.M., Baker, E.S.: Uncovering PFAS and other xenobiotics in the dark metabolome using ion mobility spectrometry, mass defect analysis, and machine learning. Environ. Sci. Technol. **56**(12), 9133–9143 (2022)
25. Abd-El-Aziz, A.S., Antonietti, M., Barner-Kowollik, C., Binder, W.H., Böker, A., Boyer, C., Buchmeiser, M.R., Cheng, S.Z., D'Agosto, F., Floudas, G., Frey, H.: The next 100 years of polymer science. Macromol. Chem. Phys. **221**(16), 2000216 (2020)

26. Zhai, C., Li, T., Shi, H., Yeo, J.: Discovery and design of soft polymeric bio-inspired materials with multiscale simulations and artificial intelligence. J. Mater. Chem. B **8**(31), 6562–6587 (2020)
27. Van Krevelen, D.W., Te Nijenhuis, K.: Properties of Polymers: their Correlation with Chemical Structure; Their Numerical Estimation and Prediction from Additive Group Contributions. Elsevier (2009)
28. Doan Tran, H., Kim, C., Chen, L., Chandrasekaran, A., Batra, R., Venkatram, S., Kamal, D., Lightstone, J.P., Gurnani, R., Shetty, P., Ramprasad, M.: Machine-learning predictions of polymer properties with polymer genome. J. Appl. Phys **128**(17) (2020)
29. Yan, C., Li, G.: The rise of machine learning in polymer discovery. Adv. Intell. Syst. **5**(4), 2200243 (2023)
30. Upadhya, R., Kosuri, S., Tamasi, M., Meyer, T.A., Atta, S., Webb, M.A., Gormley, A.J.: Automation and data-driven design of polymer therapeutics. Adv. Drug Deliv. Rev. **171**, 1–28 (2021)
31. Sharma, A., Mukhopadhyay, T., Rangappa, S.M., Siengchin, S., Kushvaha, V.: Advances in computational intelligence of polymer composite materials: machine learning assisted modeling, analysis and design. Arch. Comput. Methods Eng. **29**(5), 3341–3385 (2022)
32. O'Reilly, C.S., Elbadawi, M., Desai, N., Gaisford, S., Basit, A.W., Orlu, M.: Machine learning and machine vision accelerate 3D printed orodispersible film development. Pharmaceutics **13**(12), 2187 (2021)
33. Bannigan, P., Bao, Z., Hickman, R.J., Aldeghi, M., Häse, F., Aspuru-Guzik, A., Allen, C.: Machine learning models to accelerate the design of polymeric long-acting injectables. Nat. Commun. **14**(1), 35 (2023)
34. Patel, R.A., Webb, M.A.: Data-driven design of polymer-based biomaterials: high-throughput simulation, experimentation, and machine learning. ACS Appl. Bio Mater. **7**(2), 510–527 (2023)
35. Day, E.C., Chittari, S.S., Bogen, M.P., Knight, A.S.: Navigating the expansive landscapes of soft materials: a user guide for high-throughput workflows. ACS Polym. Au **3**(6), 406–427 (2023)
36. Rubens, M., Van Herck, J., Junkers, T.: Automated polymer synthesis platform for integrated conversion targeting based on inline benchtop NMR. ACS Macro Lett. **8**(11), 1437–1441 (2019)
37. Basak, S., Bandyopadhyay, A.: Toward making polymer chemistry autonomous. ACS Appl. Eng. Mater. **2**(5), 1190–1208 (2024)
38. Sarker, I.H.: AI-based modeling: techniques, applications and research issues towards automation, intelligent and smart systems. SN Comput. Sci. **3**(2), 158 (2022)
39. Bewersdorff, X.Z., Roberts, J., Nerdel, C.: Comput. Educ.: Artif. Intell. **4**, 100143 (2023)
40. Coveney, P.V., Highfield, R.: Artificial intelligence must be made more scientific. J. Chem. Inf. Model. **64**(15), 5739–5741 (2024)
41. Agrawal, A., Gans, J., Goldfarb, A.: Prediction, judgment, and complexity: a theory of decision-making and artificial intelligence. In: The Economics of Artificial Intelligence: an Agenda, pp. 89–110. University of Chicago Press (2018)
42. Duan, Y., Edwards, J.S., Dwivedi, Y.K.: Artificial intelligence for decision making in the era of big data–evolution, challenges and research agenda. Int. J. Inf. Manag. **48**, 63–71 (2019)
43. Krenn, M., Pollice, R., Guo, S.Y., Aldeghi, M., Cervera-Lierta, A., Friederich, P., dos Passos Gomes, G., Häse, F., Jinich, A., Nigam, A., Yao, Z.: On scientific understanding with artificial intelligence. Nat. Rev. Phys. **4**(12), 761–769 (2022)
44. Hassija, V., Chamola, V., Mahapatra, A., Singal, A., Goel, D., Huang, K., Scardapane, S., Spinelli, I., Mahmud, M., Hussain, A.: Interpreting black-box models: a review on explainable artificial intelligence. Cognit. Comput. **16**, 45–74 (2024)
45. Wang, A.Y.T., Murdock, R.J., Kauwe, S.K., Oliynyk, A.O., Gurlo, A., Brgoch, J., Persson, K.A., Sparks, T.D.: Machine learning for materials scientists: an introductory guide toward best practices. Chem. Mater. **32**(12), 4954–4965 (2020)

46. Yang, Z., Wu, J.G., Xie, H.: Taming Frankenstein's monster: Ethical considerations relating to generative artificial intelligence in education. Asia Pacific J. Educ. 1–14 (2024)
47. Shkodinsky, S., Syrovatskaya, V., Khokhoeva, Z.: The role of artificial intelligence in forecasting and managing technical risks. Reliabil.: Theory Appl. **19**(SI 6 (81)), 210–216 (2024)
48. van den Hurk, R.S., Pirok, B.W., Bos, T.S.: The role of artificial intelligence and machine learning in polymer characterization: emerging trends and perspectives. Chromatographia 1–7 (2025)
49. Olivetti, E.A., Cole, J.M., Kim, E., Kononova, O., Ceder, G., Han, T.Y.J., Hiszpanski, A.M.: Data-driven materials research enabled by natural language processing and information extraction. Appl. Phys. Rev. **7**(4) (2020)
50. O'Gorman, T., Jensen, Z., Mysore, S., Huang, K., Mahbub, R., Olivetti, E., McCallum, A.: MS-mentions: consistently annotating entity mentions in materials science procedural text. In: Proceedings of the 2021 Conference on Empirical Methods in Natural Language Processing, pp. 1337–1352 (2021)
51. Fan, V., Qian, Y., Wang, A., Wang, A., Coley, C.W., Barzilay, R.: OpenChemIE: an information extraction toolkit for chemistry literature. J. Chem. Inf. Model. **64**(14), 5521–5534 (2024)
52. Shetty, P., Rajan, A.C., Kuenneth, C., Gupta, S., Panchumarti, L.P., Holm, L., Zhang, C., Ramprasad, R.: A general-purpose material property data extraction pipeline from large polymer corpora using natural language processing. NPJ Comput. Mater. **9**(1), 52 (2023)
53. Swarnkar, S.K., Patra, J.P., Kshatri, S.S., Rathore, Y.K., Tran, T.A., (eds.): Supervised and Unsupervised Data Engineering for Multimedia Data. Wiley (2024)
54. Berman, H.M., Westbrook, J., Feng, Z., Gilliland, G., Bhat, T.N., Weissig, H., Shindyalov, I.N., Bourne, P.E.: The protein data bank. Nucleic Acids Res. **28**(1), 235–242 (2000)
55. Kim, S., Chen, J., Cheng, T., Gindulyte, A., He, J., He, S., Li, Q., Shoemaker, B.A., Thiessen, P.A., Yu, B., Zaslavsky, L.: PubChem 2023 update. Nucleic Acids Res. **51**(D1), D1373–D1380 (2023)
56. Pence, H.E., Williams, A.: ChemSpider: an online chemical information resource (2010)
57. Ruddigkeit, L., Van Deursen, R., Blum, L.C., Reymond, J.L.: Enumeration of 166 billion organic small molecules in the chemical universe database GDB-17. J. Chem. Inf. Model. **52**(11), 2864–2875 (2012)
58. Ramakrishnan, R., Dral, P.O., Rupp, M., Von Lilienfeld, O.A.: Quantum chemistry structures and properties of 134 kilo molecules. Sci. Data **1**(1), 1–7 (2014)
59. Otsuka, S., Kuwajima, I., Hosoya, J., Xu, Y., Yamazaki, M.: PoLyInfo: polymer database for polymeric materials design. In: 2011 International Conference on Emerging Intelligent Data and Web Technologies, pp. 22–29. IEEE (2011)
60. Dong, G., Liu, H., (eds.): Feature Engineering for Machine Learning and Data Analytics. CRC press (2018)
61. Carraher Jr, C.E., Seymour, R.B.: Structure—Property Relationships in Polymers. Springer Science & Business Media (2012)
62. Sultan, H.M., Menesy, A.S., Kamel, S., Turky, R.A., Hasanien, H.M., Al-Durra, A.: Optimal values of unknown parameters of polymer electrolyte membrane fuel cells using improved chaotic electromagnetic field optimization. IEEE Trans. Ind. Appl. **57**(6), 6669–6687 (2021)
63. Wolpert, D.H., Macready, W.G.: No free lunch theorems for optimization. IEEE Trans. Evol. Comput. **1**(1), 67–82 (1997)
64. Bishop, C.M.: Model-based machine learning. Philos. Trans. R. Soc. A: Math. Phys. Eng. Sci. **371**(1984), 20120222 (2013)
65. Xu, P., Chen, H., Li, M., Lu, W.: New opportunity: machine learning for polymer materials design and discovery. Adv. Theory Simul. **5**(5), 2100565 (2022)
66. Morales, E.F., Escalante, H.J.: A brief introduction to supervised, unsupervised, and reinforcement learning. In: Biosignal Processing and Classification using Computational Learning and Intelligence, pp. 111–129. Academic Press (2022)
67. Chinnamgari, S.K.: R Machine Learning Projects: implement Supervised, Unsupervised, and Reinforcement Learning Techniques using R 3.5. Packt Publishing Ltd. (2019)

68. Nasteski, V.: An overview of the supervised machine learning methods. Horizons. B **4**(51–62), 56 (2017)
69. Huang, J., Sun, Y., Zhang, J.: Reduction of computational error by optimizing SVR kernel coefficients to simulate concrete compressive strength through the use of a human learning optimization algorithm. Eng. Comput. 1–18 (2022)
70. Salman, R., Kecman, V.: Regression as classification. In: 2012 Proceedings of IEEE Southeastcon, pp. 1–6. IEEE (2012)
71. Elsayad, A.M., Zeghid, M., Ahmed, H.Y., Elsayad, K.A.: Exploration of biodegradable substances using machine learning techniques. Sustainability **15**(17), 12764 (2023)
72. Lin, C., Zhang, H.: Polymer biodegradation in aquatic environments: a machine learning model informed by meta-analysis of structure-biodegradation relationships. Environ. Sci. Technol. (2025)
73. Liu, T.L., Liu, L.Y., Ding, F., Li, Y.Q.: A machine learning study of polymer-solvent interactions. Chin. J. Polym. Sci. **40**(7), 834–842 (2022)
74. Das, D.K., Ghosh, M., Pal, M., Maiti, A.K., Chakraborty, C.: Machine learning approach for automated screening of malaria parasite using light microscopic images. Micron **45**, 97–106 (2013)
75. Qu, E.Z., Jimenez, A.M., Kumar, S.K., Zhang, K.: Quantifying nanoparticle assembly states in a polymer matrix through deep learning. Macromolecules **54**(7), 3034–3040 (2021)
76. Ren, Z., Wang, S., Zhang, Y.: Weakly supervised machine learning. CAAI Trans. Intell. Technol. **8**(3), 549–580 (2023)
77. Malik, A., Tuckfield, B.: Applied Unsupervised Learning with R: uncover Hidden Relationships and Patterns with k-Means Clustering, Hierarchical Clustering, and PCA. Packt Publishing Ltd. (2019)
78. Swann, E., Sun, B., Cleland, D.M., Barnard, A.S.: Representing molecular and materials data for unsupervised machine learning. Mol. Simul. **44**(11), 905–920 (2018)
79. Saxena, A., Prasad, M., Gupta, A., Bharill, N., Patel, O.P., Tiwari, A., Er, M.J., Ding, W., Lin, C.T.: A review of clustering techniques and developments. Neurocomputing **267**, 664–681 (2017)
80. Solorio-Fernández, S., Carrasco-Ochoa, J.A., Martínez-Trinidad, J.F.: A review of unsupervised feature selection methods. Artif. Intell. Rev. **53**(2), 907–948 (2020)
81. Ziolek, R.M., Smith, P., Pink, D.L., Dreiss, C.A., Lorenz, C.D.: Unsupervised learning unravels the structure of four-arm and linear block copolymer micelles. Macromolecules **54**(8), 3755–3768 (2021)
82. Zhao, W., Huo, H., Sun, Z.Y., Lu, Z.Y.: Data-driven analysis of dynamical heterogeneity in polymer melts near surfaces. Comput. Mater. Sci. **235**, 112811 (2024)
83. Botvinick, M., Ritter, S., Wang, J.X., Kurth-Nelson, Z., Blundell, C., Hassabis, D.: Reinforcement learning, fast and slow. Trends Cogn. Sci. **23**(5), 408–422 (2019)
84. Venugopal, G.: Exploring reinforcement learning models for various aspects of decision making (Doctoral dissertation, International Institute of Information Technology Hyderabad) (2024)
85. Singh, A., Yu, A., Yang, J., Zhang, J., Kumar, A., Levine, S.: Cog: connecting new skills to past experience with offline reinforcement learning (2020). arXiv:2010.14500
86. Zangirolami, V., Borrotti, M.: Dealing with uncertainty: Balancing exploration and exploitation in deep recurrent reinforcement learning. Knowl.-Based Syst. **293**, 111663 (2024)
87. Li, H., Collins, C.R., Ribelli, T.G., Matyjaszewski, K., Gordon, G.J., Kowalewski, T., Yaron, D.J.: Tuning the molecular weight distribution from atom transfer radical polymerization using deep reinforcement learning. Mol. Syst. Des. Eng. **3**(3), 496–508 (2018)
88. Ma, R., Zhang, H., Luo, T.: Exploring high thermal conductivity amorphous polymers using reinforcement learning. ACS Appl. Mater. Interfaces **14**(13), 15587–15598 (2022)
89. Knox, S.T., Parkinson, S.J., Wilding, C.Y., Bourne, R.A., Warren, N.J.: Autonomous polymer synthesis delivered by multi-objective closed-loop optimisation. Polym. Chem. **13**(11), 1576–1585 (2022)

90. Beaucage, P.A., Sutherland, D.R., Martin, T.B.: Automation and machine learning for accelerated polymer characterization and development: past, potential, and a path forward. Macromolecules **57**(18), 8661–8670 (2024)
91. Cutler, M., Walsh, T.J., How, J.P.: Real-world reinforcement learning via multifidelity simulators. IEEE Trans. Rob. **31**(3), 655–671 (2015)
92. Tao, L., Varshney, V., Li, Y.: Benchmarking machine learning models for polymer informatics: an example of glass transition temperature. J. Chem. Inf. Model. **61**(11), 5395–5413 (2021)
93. Ryzhkov, F.V., Ryzhkova, Y.E., Elinson, M.N.: Python tools for structural tasks in chemistry. Mol. Div. 1–20 (2024)
94. Yan, C., Feng, X., Wick, C., Peters, A., Li, G.: Machine learning assisted discovery of new thermoset shape memory polymers based on a small training dataset. Polymer **214**, 123351 (2021)
95. Marquez, R., Barrios, N., Vera, R.E., Mendez, M.E., Tolosa, L., Zambrano, F., Li, Y.: A perspective on the synergistic potential of artificial intelligence and product-based learning strategies in biobased materials education. Educ. Chem. Eng. **44**, 164–180 (2023)
96. Zhang, H., He, Q., Zhang, F., Duan, Y., Qin, M., Feng, W.: Biomimetic intelligent thermal management materials: from nature-inspired design to machine-learning-driven discovery. Adv. Mater. 2503140 (2025)
97. Shahzad, K., Mardare, A.I., Hassel, A.W.: Accelerating materials discovery: combinatorial synthesis, high-throughput characterization, and computational advances. Sci. Technol. Adv. Mater.: Methods **4**(1), 2292486 (2024)
98. Maqsood, A., Chen, C., Jacobsson, T.J.: The future of material scientists in an age of artificial intelligence. Adv. Sci. **11**(19), 2401401 (2024)
99. Hornik, K., Stinchcombe, M., White, H.: Multilayer feedforward networks are universal approximators. Neural Netw. **2**(5), 359–366 (1989)
100. Raccuglia, P., Elbert, K.C., Adler, P.D., Falk, C., Wenny, M.B., Mollo, A., Zeller, M., Friedler, S.A., Schrier, J., Norquist, A.J.: Machine-learning-assisted materials discovery using failed experiments. Nature **533**(7601), 73–76 (2016)
101. Boobier, S., Hose, D.R., Blacker, A.J., Nguyen, B.N.: Machine learning with physicochemical relationships: solubility prediction in organic solvents and water. Nat. Commun. **11**(1), 5753 (2020)
102. Wu, Y., Wang, G.: Machine learning based toxicity prediction: from chemical structural description to transcriptome analysis. Int. J. Mol. Sci. **19**(8), 2358 (2018)
103. Jukič, M., Bren, U.: Machine learning in antibacterial drug design. Front. Pharmacol. **13**, 864412 (2022)
104. She, C., Huang, Q., Chen, C., Jiang, Y., Fan, Z., Gao, J.: Machine learning-guided search for high-efficiency perovskite solar cells with doped electron transport layers. J. Mater. Chem. A **9**(44), 25168–25177 (2021)
105. Liu, Y., Yan, W., Han, S., Zhu, H., Tu, Y., Guan, L., Tan, X.: How machine learning predicts and explains the performance of perovskite solar cells. Solar RRL **6**(6), 2101100 (2022)
106. Rajan, K.: Materials informatics. Mater. Today **8**(10), 38–45 (2005)
107. Schmidt, J., Marques, M.R., Botti, S., Marques, M.A.: Recent advances and applications of machine learning in solid-state materials science. NPJ Comput. Mater. **5**(1), 83 (2019)
108. Agrawal, A., Choudhary, A.: Perspective: Materials informatics and big data: Realization of the "fourth paradigm" of science in materials science. Appl. Mater. **4**(5) (2016)
109. Baker, N., Alexander, F., Bremer, T., Hagberg, A., Kevrekidis, Y., Najm, H., Parashar, M., Patra, A., Sethian, J., Wild, S., Willcox, K.: Workshop report on basic research needs for scientific machine learning: core technologies for artificial intelligence. USDOE Office of Science (SC), Washington, DC (United States) (2019)
110. Lookman, T., Alexander, F.J., Bishop, A.R.: Perspective: Codesign for materials science: an optimal learning approach. APL Mater. **4**(5) (2016)
111. Ren, F., Ward, L., Williams, T., Laws, K. J., Wolverton, C., Hattrick-Simpers, J., Mehta, A.: Accelerated discovery of metallic glasses through iteration of machine learning and high-throughput experiments. Sci. Adv. **4**(4), eaaq1566 (2018)

112. Kusne, A.G., Gao, T., Mehta, A., Ke, L., Nguyen, M.C., Ho, K.M., Antropov, V., Wang, C.Z., Kramer, M.J., Long, C., Takeuchi, I.: On-the-fly machine-learning for high-throughput experiments: search for rare-earth-free permanent magnets. Sci. Rep. **4**(1), 6367 (2014)
113. Suram, S.K., Xue, Y., Bai, J., Le Bras, R., Rappazzo, B., Bernstein, R., Bjorck, J., Zhou, L., Van Dover, R.B., Gomes, C.P., Gregoire, J.M.: Automated phase mapping with AgileFD and its application to light absorber discovery in the V-Mn–Nb oxide system. ACS Comb. Sci. **19**(1), 37–46 (2017)
114. Dunn, A., Brenneck, J., Jain, A.: Rocketsled: a software library for optimizing high-throughput computational searches. J. Phys.: Mater. **2**(3), 034002 (2019)
115. Gongora, A.E., Xu, B., Perry, W., Okoye, C., Riley, P., Reyes, K.G., Morgan, E.F., Brown, K.A.: A Bayesian experimental autonomous researcher for mechanical design. Sci. Adv. **6**(15), eaaz1708 (2020)
116. Reddy, C.K., Shojaee, P.: Towards scientific discovery with generative AI: progress, opportunities, and challenges. In: Proceedings of the AAAI Conference on Artificial Intelligence, vol. 39, no. 27, pp. 28601–28609 (2020)
117. DeCost, B.L., Hattrick-Simpers, J.R., Trautt, Z., Kusne, A.G., Campo, E., Green, M.L.: Scientific AI in materials science: a path to a sustainable and scalable paradigm. Mach. Learn.: Sci. Technol. **1**(3), 033001 (2020)
118. Jia, X., Lynch, A., Huang, Y., Danielson, M., Lang'at, I., Milder, A., Ruby, A.E., Wang, H., Friedler, S.A., Norquist, A.J., Schrier, J.: Anthropogenic biases in chemical reaction data hinder exploratory inorganic synthesis. Nature **573**(7773), 251–255 (2019)
119. Wang, H., Fu, T., Du, Y., Gao, W., Huang, K., Liu, Z., Chandak, P., Liu, S., Van Katwyk, P., Deac, A., Anandkumar, A.: Scientific discovery in the age of artificial intelligence. Nature **620**(7972), 47–60 (2023)
120. Waly, M.F.: Artificial intelligence and scientific research. Sustain. Educ. Globe **2**(1), 1–14 (2024)
121. Youvan, D.: The role of AI in revolutionizing academic publishing and knowledge dissemination (2024)
122. Pyzer-Knapp, E.O., Pitera, J.W., Staar, P.W., Takeda, S., Laino, T., Sanders, D.P., Sexton, J., Smith, J.R., Curioni, A.: Accelerating materials discovery using artificial intelligence, high performance computing and robotics. NPJ Comput. Mater. **8**(1), 84 (2022)
123. Chen, C., Maqsood, A., Jacobsson, T.J.: The role of machine learning in perovskite solar cell research. J. Alloy. Compd. **960**, 170824 (2023)
124. Westermayr, J., Gastegger, M., Schütt, K.T., Maurer, R.J.: Perspective on integrating machine learning into computational chemistry and materials science. J. Chem. Phys. **154**(23) (2021)
125. Kalidindi, S.R., De Graef, M.: Materials data science: current status and future outlook. Annu. Rev. Mater. Res. **45**(1), 171–193 (2015)
126. Draxl, C., Scheffler, M.: Big data-driven materials science and its FAIR data infrastructure. In: Handbook of Materials Modeling, pp. 1–25. Springer, Cham (2019)
127. Jain, A., Ong, S.P.: New opportunities for materials informatics: resources and data mining techniques for uncovering hidden relationships. APL Mater. **1**, 011002 (2013)
128. Curtarolo, S., Setyawan, W., Hart, G.L., Jahnatek, M., Chepulskii, R.V., Taylor, R.H., Wang, S., Xue, J., Yang, K., Levy, O., Mehl, M.J., Morgan, D.: AFLOW: an automatic framework for high-throughput materials discovery. Comput. Mater. Sci. **58**, 218–226 (2012)
129. He, K., Zhang, X., Ren, S., Sun, J.: Deep residual learning for image recognition. In: Proceedings of the IEEE Conference on Computer Vision and Pattern Recognition, pp. 770–778 (2016)
130. Stier, S.P., Kreisbeck, C., Ihssen, H., Popp, M.A., Hauch, J., Malek, K., Reynaud, M., Goumans, T.P.M., Carlsson, J., Todorov, I., Gold, L.: Materials acceleration platforms (MAPs): accelerating materials research and development to meet urgent societal challenges. Adv. Mater. **36**(45), 2407791 (2024)
131. Alekseeva, L., Azar, J., Giné, M., Samila, S., Taska, B.: The demand for AI skills in the labor market. Labour Econ. **71**, 102002 (2021)

132. Boyce, B., Dingreville, R., Desai, S., Walker, E., Shilt, T., Bassett, K.L., Wixom, R.R., Stebner, A.P., Arroyave, R., Hattrick-Simpers, J., Warren, J.A.: Machine learning for materials science: barriers to broader adoption. Matter **6**(5), 1320–1323 (2023)
133. Kalidindi, S.: Materials data sciences and informatics. Coursera [Online course] (2020)
134. Gastelum, J.C.V., Strachan, A., Desai, S.: Machine learning for materials science: part 1 (2019)
135. Rajashekar, N.C., Shin, Y.E., Pu, Y., Chung, S., You, K., Giuffre, M., Chan, C.E., Saarinen, T., Hsiao, A., Sekhon, J., Wong, A.H.: Human-algorithmic interaction using a large language model-augmented artificial intelligence clinical decision support system. In: Proceedings of the 2024 CHI Conference on Human Factors in Computing Systems, pp. 1–20 (2024)
136. Himanen, L., Geurts, A., Foster, A.S., Rinke, P.: Data-driven materials science: status, challenges, and perspectives. Adv. Sci. **6**(21), 1900808 (2019)
137. Annapareddy, V. N.: Connected Intelligence: transforming Education and Energy with Big Data, Cloud Connectors, and Artificial Intelligence. Deep Science Publishing (2025)
138. Olafare, F.O.: Artificial intelligence in educational technology: Panacea to efficient instructional delivery. In: Educational Broadcasting in Nigeria in the Age of Artificial Intelligence, pp. 217–223. Springer Nature Switzerland, Cham (2024)
139. Madhu, R.: Ensuring Equity and AI, Personalization, Equity, and the Future of Learning, p. 93 (2025)
140. Voudoukis, N., Pagiatakis, G.: Massive open online courses (MOOCs): practices, trends, and challenges for the higher education. Eur. J. Educ. Pedagogy **3**(3), 288–295 (2022)
141. Yu, H., Miao, C., Leung, C., White, T.J.: Towards AI-powered personalization in MOOC learning. NPJ Sci. Learn. **2**(1), 15 (2017)
142. Mahamad, S., Chin, Y.H., Zulmuksah, N.I.N., Haque, M.M., Shaheen, M., Nisar, K.: Technical review: architecting an AI-driven decision support system for enhanced online learning and assessment. Future Internet **17**(9), 383 (2025)
143. Gligorea, I., Cioca, M., Oancea, R., Gorski, A.T., Gorski, H., Tudorache, P.: Adaptive learning using artificial intelligence in e-learning: a literature review. Educ. Sci. **13**(12), 1216 (2023)
144. Koka, N.A.: An insight into the efficiency of artificial intelligence (AI)-chatbot as digital tutors for enhancing learners' motivation and performance in linguistics courses. Pakistan J. Life Soc. Sci. **22**(2) (2024)
145. Ogwueleka, F.N.: Plagiarism detection in the age of artificial intelligence: current technologies and future directions. In: AI and Ethics, Academic Integrity and the Future of Quality Assurance in Higher Education, p. 10 (2025)
146. Martha, A.S.D., Widowati, S., Rahayu, D.P., Nursyawal, M.I., Hariz, R.R.: Examining usability and student experience of ChatGPT as a learning tool on Moodle in higher education. In: 2024 International Conference on Information Technology Systems and Innovation (ICITSI), pp. 8–13. IEEE (2024)
147. Ashadevi, B., Muthamil Selvi, P.: Open educational resources in computer science: opportunities and challenges. Int. J. Comput. Sci. Issues **14**(5), 42–49 (2017)
148. Hussein, R.M., Hassan, B.A.: Collaboration tools and their role in agile software projects (2025). arXiv:2506.10985
149. Ouyang, F., Zheng, L., Jiao, P.: Artificial intelligence in online higher education: a systematic review of empirical research from 2011 to 2020. Educ. Inf. Technol. **27**(6), 7893–7925 (2022)
150. Brandl, K., Rabadia, S.V., Chang, A., Mandel, J.: Benefits of focus group discussions beyond online surveys in course evaluations by medical students in the United States: a qualitative study. J. Educ. Evaluat. Health Prof. **15** (2018)
151. Wu, C., Zipf, S., Li, N., Hellar, D.B.: Data-Informed instruction: pedagogical responses and obstacles in using learning analytics. In: 2025 ASEE Annual Conference & Exposition (2025)
152. Delgado Algarra, E.J., Bernal Bravo, C., Morales Cevallos, M.B., López Meneses, E.: Pedagogical and technical analyses of massive open online courses on artificial intelligence. Appl. Sci. **14**(3), 1051 (2024)
153. Janahi, Y., Obeidat, M.: Predictive analytics and AI in education: a systematic literature review on identifying and supporting at-risk students (2025). SSRN 5240920

MIX
Papier aus verantwortungsvollen Quellen
Paper from responsible sources
FSC® C105338

If you have any concerns about our products,
you can contact us on
ProductSafety@springernature.com

In case Publisher is established outside the EU,
the EU authorized representative is:
Springer Nature Customer Service Center GmbH
Europaplatz 3, 69115 Heidelberg, Germany

Printed by Libri Plureos GmbH
in Hamburg, Germany